Advanced Wireless LAN

Advanced Wireless LAN

Editor

Dan Hallmen

scitus
academics

Advanced Wireless LAN

Edited by **Dan Hallmen**

ISBN: 978-1-68117-279-8
Library of Congress Control Number: 2016937451

SCITUS Academics LLC,
www.scitusacademics.com
Box No. 4766, 616 Corporate Way,
Suite 2, Valley Cottage,
NY 10989

Notice

Printed in United States of America on Acid Free Paper

Preface

A local wireless network (WLAN) is a wireless network connecting two or more devices to a limited area, for example, a wireless distribution system used for home, school, computer lab or office building. This allows users to move into a local coverage area and to connect to the network and the wider Internet. Most modern WLANs are based on the IEEE 802.11 standards, marketed under the Wi-Fi brand. Wireless LANs have become popular at home and in commercial complexes offering wireless access to their customers due to easy installation and use. The fundamental structure of all networks consists of a main computer or server and connected clients. The server typically has two network interface cards (NICs) installed and software for network support. Since Windows98 Special Edition (SE), Microsoft Windows operating systems have built-in networking capabilities, but third party networking software is also available. A desktop could be the server while a laptop could be a simple home wireless LAN client. In the office, a wireless LAN provides instant connectivity for mobile personnel. It also avoids the costly costs of running. Ethernet cables across a building and easily connects customers to the desktop. Since cabling for customers is not available, easy installation is one of the main advantages of a wireless LAN. Wireless LANs are also more flexible and cheaper than wired LANs.

This book offers state-of-the-art information in the field of wireless LAN. It offers both practitioners and researchers the opportunity to discover the problems arising from the rapidly evolving LAN wireless technologies.

Table of Contents

1

CHAPTER

ESTIMATING THE NUMBER OF COMPETING TERMINALS WITHOUT A STATE VARIATION DETECTOR IN WIRELESS LAN

Jaechan Lim, Taejin Kim and Daehyoung Hong

ABSTRACT

Estimating the number of competing terminals n (who wish to transmit a packet at the same time) in the IEEE 802.11 system is important for system throughput performance because optimal back-off window size needs to be selected based on n. Therefore, as a new approach for estimating n, we propose H infinity filter that does not need a state variation detector as opposed to the cases of previously proposed approaches. The state variation detector's flaw is incurring tracking latency in addition to the side effect of increased computational cost. All previously proposed approaches demand the employment of the state variation detector to detect the variation of n in the IEEE 802.11 system. By employing H infinity filter, we show improved throughput performance of the system compared to that of previously proposed approaches (e.g., the Kalman filter and particle filter) based on the improved performance in tracking n. In this paper, we justify the superiority of the proposed approach in the terms of tracking performance, throughput performance, and computational complexity.

Keywords Competing terminals Gaussian mixture sigma point particle filter H infinity filter IEEE 802.11 Kalman filter

1 INTRODUCTION

IEEE 802.11 wireless local area network (WLAN) protocol [1] is commonly employed in response to the increased usage of multi-protocol mobile devices such as smartphones. This protocol shares medium resources based on distributed coordination function (DCF) in its medium access control (MAC) layer. DCF is a random access scheme and employs 'carrier sense multiple access with the collision avoidance (CSMA/CA)' protocol [2, 3]. If a terminal detects a collision with other terminals for medium access, it waits for a random time interval and retries to acquire medium resources according to the protocol. The terminal selects random time interval within 'binary exponential back-off window size.' The throughput performance of the system can be maximized by selecting the optimal window size based on the number of competing terminals (n) which try to transmit packets simultaneously [4–6]. According to the DCF protocol, a terminal can choose either the 'basic access mode' or 'request-to-send/clear-to-send (RTS/CTS) access mode' depending on a predetermined RTS threshold of the packet payload length. It has been shown in [7] that the RTS threshold which maximizes the system throughput performance also depends on n[8]. Therefore, there are a number of motivations for the accurate estimation of n; nonetheless, the task is not easy because n varies without any prior information nor statistical knowledge.

The extended Kalman filter (EKF) approach was proposed in [9] where a 'state variation detector based on cumulative summary (CUSUM) algorithm' is additionally employed to adaptively change the state noise variance when the variation of n is detected. Particle filtering (PF) was also proposed in [10] to combat the non-Gaussian/non-linear problem. Recently, Gaussian mixture sigma point particle filtering (GMSPPF) was proposed in [11] that shows outperforming results over existing methods, e.g., the EKF, unscented Kalman filter (UKF), PF, unscented PF (UPF).

The novel contribution of this paper can be specified in the following. In this paper, we propose a new approach, i.e., H infinity filter (HIF) for estimating n, by which we can obtain enhanced throughput performance compared to that obtained by previously proposed approaches. The proposed HIF can be employed without the knowledge on noise statistics of the problem (i.e., the state and measurement noise). Consequently, we do not need to know the variance of the state noise as opposed to the cases of the EKF and PF approaches. In the usage of the EKF and PF, we need the knowledge on the mean and variance of the state and measurement noise; furthermore, Gaussian noise is assumed in the applications of the EKF. However, in the investigated problem of this paper, the varying pattern of the state (i.e. n) is hard to be known. Therefore, the state variation detector (i.e. CUSUM algorithm) was additionally employed in the application of the EKF [9]. This is the same case for the application of PF to the investigated problem. In other words, additional CUSUM algorithm needs to be employed for the application of PF too for this problem. This state variation detection algorithm also causes the delay in the tracking ability against the prompt detection of the state variation. Consequently, the employment of HIF requires the least computational complexity due to the non-necessity of the additional CUSUM algorithm for the detection of the state variation. The computational complexity of PF is proportional to the number of particles (e.g., if 100 particles are employed, the computational complexity of PF is approximately 100 times as that of the EKF or HIF). Therefore, the advantages of the proposed filter are the superiority in the computational complexity and prompt tracking ability that results in the better mean squared error performance for accurate tracking of n. Furthermore, accurate tracking performance affects the throughput performance of the network. These advantages originally result from the feature of the HIF that we can apply this filter with unknown noise statistics (both state and measurement). In this paper, we justify the superiority and advantageous features of the proposed HIF in terms of tracking ability, throughput performance, and computational complexity.

2 THE PROBLEM FORMULATION

2.1 Estimation in dynamic state system

Many problems in the area of signal processing, a parameter of interest is estimated based on some measurement. Particularly, if the system model can be described by the discrete time-varying states with corresponding measurement, we can estimate the state sequentially by applying dynamic filters. The dynamic state system that $\boldsymbol{n}$ and observed measurement $\boldsymbol{\xi}$ with zero mean and additive noise processes of $\mathbf{w}$ and $\mathbf{v}$ at time step k is expressed as follows:

$$\boldsymbol{n}_k = g(\boldsymbol{n}_{k-1}) + \boldsymbol{w}_k, \tag{1}$$

$$\boldsymbol{\xi}_k = h(\boldsymbol{n}_k) + \boldsymbol{v}_k, \tag{2}$$

where boldface denotes a vector, and $g(\cdot)$ and $h(\cdot)$ are the given state transition and the observation function, respectively, which are possibly non-linear. According to this system model, we can estimate the time-varying state $\mathbf{n}_k$ sequentially based on the observed measurement ξ_k by dynamic filters, e.g., the Kalman filter, particle filter, and their variants.

In general cases, i.e., apart from the linear and Gaussian case, there exist a number of sub-optimal approaches such as particle filtering (PF) [12, 13]. The Kalman filter also can be applied as a sub-optimal approach to non-linear model by extending it using Taylor series (in this case, still Gaussian noise is assumed) [14].

There are also variants of the EKF and those of PF approaches, e.g., UKF, Gaussian PF, auxiliary PF, likelihood PF, etc. Although PF approaches require high computational cost, they outperform EKF and its variants in most highly non-linear problems; furthermore, the Gaussian noise does not have to be assumed.

2.2 Dynamic state system for estimating *n*

All the filtering methods, i.e. the proposed approach in this paper and previously proposed approaches, basically estimate the dynamically varying (in time) state by using a measurement which is a function of the state. Therefore, the problem of estimating n needs to be modeled by the 'dynamic state system' and consequent 'measurement equation' first.

Following [7, 9], we consider a scenario of the fixed number of n competing terminals. Each terminal operates in the saturated network and ideal channel conditions: the saturated network condition means that all terminals always have a packet waiting for transmission, and neither packet corruption nor hidden terminals are assumed in the ideal channel condition. ξ denotes the probability that a packet is transmitted on a collided channel. Then, n can be expressed as the function of ξ as follows [9].

$$n = f(\xi) = 1 + \frac{\log(1-\xi)}{\log\left(1 - \frac{2(1-2\xi)}{(1-2\xi)(D_{\min}+1)+\xi D_{\min}(1-(2\xi)^m)}\right)}, \tag{3}$$

where m and $D_{\min}$ denote the maximum back-off stage and the minimum contention window sizes, respectively. After a collision, each terminal doubles its contention window size up to the maximum value $D_{\max}=2^m D_{\min}$, where $m=\log2(D_{\max}/D_{\min})$. Maximization of the saturation throughput of IEEE 802.11 systems has been studied [4, 7]. Saturation throughput is defined as 'the number of successfully transmitted payload slots by all terminals over observed slot numbers.' It can be thought of as the channel utilization efficiency. In order to maximize the saturation throughput, the minimum back-off window size is properly selected, and the optimal relationship is analytically obtained as follows:

$$D_{\min} = n\sqrt{2T},$$

(4)

where T is the packet transmission time that comprises the header, SIFS, ACK, and DIFS. The size of T depends on the employed PHY and transmitted payload size; therefore, it can be easily obtained to the terminal. $D_{\min}$ needs to be selected according to (4) to obtain optimal saturation throughput of the system. From (3), we can estimate the number of competing terminals by measuring collision probability ξregardless of the packet size. ξ is measured by each terminal at the time step k (each time step comprises B slots) as follows:

$$\xi_k = \frac{1}{B} \sum_{i=(k-1)B}^{kB-1} C_i,$$

(5)

where C_i=0 if the corresponding slot is empty or the transmission is successful during the slot i while C_i=1 if the slot is busy or collided.

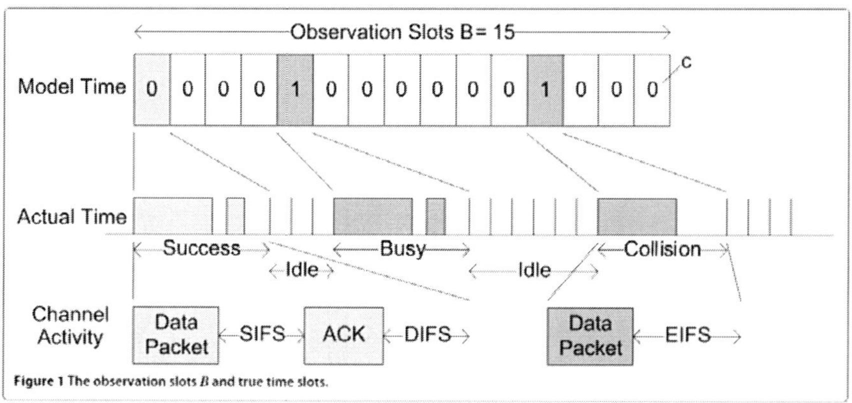

Figure 1. The observation slots *B* and true time slots.

B is a constant value; however, it does not mean that the true observation time duration is constant [9, 15]. As shown in Figure 1, one slot can have the duration of an empty slot (defined by the standard) or last the transmission time of either a successful transmission or a collision. If the packet length or n increases, then the actual observation time increases even if B remains the same. This will degrade the prompt tracking performance.

The discrete time dynamic state system model for the number of competing terminals is defined as follows [9].

$$n_k = n_{k-1} + w_k, \tag{6}$$

where w_k is the state process noise.

The obtained measurement (5) is related with the state n_k based on (3) as follows:

$$\xi_k = f^{-1}(n_k) + v_k = h(n_k) + v_k, \tag{7}$$

where $h(\cdot)$ is the inverse function of (3), v_k is the observation noise that follows the binomial distribution with zero mean, and the variance can be computed as follows [15]:

$$\mathrm{Var}[v_k] = \frac{h(n_k)\cdot[1 - h(n_k)]}{B}. \tag{8}$$

Therefore, the system model of the problem is non-linear and non-Gaussian. Based on the problem formulation, i.e., (6) and (7), we are ready to apply all the filtering approaches now. It is assumed that (6) and (7) are known except for the information of w_k. The state variation detector is employed to obtain even a limited information of the statistics of w_k in the previously proposed approaches, while preliminary training process with unknown statistics is performed in the extended HIF (EHIF) approach. The filtering methods estimate n_k by using the measured information of (7) at every time step. Note that all variables (n_k, ξ_k, w_k, v_k) are expressed in a scalar form because the state and measurement are all scalar in this problem.

3 FILTERING METHODS

In this section, we propose H infinity filter for the problem of estimating the number of competing terminals and also discuss previously proposed approaches.

3.1 Previously proposed approaches

The Kalman filter is the optimal method if the problem is linear and jointly Gaussian. The investigated problem is non-linear and non-Gaussian. The EKF can be employed in non-linear systems using the Taylor expansion. This EKF algorithm was proposed in [9], and additional CUSUM algorithm was employed for the state variation detector. Using the CUSUM algorithm, the EKF changes the state noise variance $Q_k$adaptively. This approach may work well, but it has considerable latency. Another disadvantage for limited performance is that only two deterministic variances are used regardless of statistical variation pattern of n.

GMSPPF [16] adopts an 'UKF-based Gaussian sum filter' in particle filtering framework. GMSPPF employs the Gaussian mixture model (GMM), and GMM has multiple means and variances to represent a complicated density which cannot be represented by a normal Gaussian model. GMSPPF uses $G{\cdot}I{\cdot}J$ numbers of parallel UKF banks (where G is the number of GMM components, I is the number of state noise components, and J is the number of measurement noise components). Furthermore, the expectation maximization algorithm is employed to obtain the GMM from particles.

In [11], GMSPPF was proposed for the investigated problem and showed that GMSPPF provides more outstanding performance compared to that of the EKF, UKF, PF, and UPF; however, this paper assessed the performance of the approaches without employing the CUSUM algorithm in any approaches, which is not appropriate. Therefore, in this paper, we apply the CUSUM algorithm to all existing approaches, and compare the performance with that of EHIF which does not need the CUSUM algorithm.

We also consider another PF approach, i.e., Gaussian PF (GPF) [17] in this paper for performance comparison although GPF has not been proposed for the investigated problem in the literature. In GPF, resampling process is replaced by another step of particle generation from a Gaussian density that has the sample mean and the sample variance of generated particles from the prior density. Due to the advantages of simplicity and low computational complexity, it is relatively easy to apply GPF algorithm among particle filtering approaches. We can expect that GPF outperforms the EKF when the CUSUM is employed for both approaches.

3.2 Proposed approach

In this paper, we propose EHIF which does not require the knowledge on noise statistics in its applications [18]. Whereas the EKF minimizes mean squared error (MSE) of the estimate, EHIF is designed to minimize the worst possible error. The original HIF was employed in control areas, and HIF has not been widely employed due to its level of mathematical understanding and the necessity of a good modeling of the system. Discrete time HIF is based on zero-sum game theory, and we minimize the cost function which is designed on the basis of the strategy that the probability of maximum expected error is minimized regardless of the opponent; therefore, the filter is a player prepared for the worst strategy that the other player (the nature) can provide, and the goal is providing a uniformly small estimation error for any processes, measurement noises, and any initial states. Consequently, we do not need to know the noise statistics of the system for the filtering process. Accordingly, the cost function in discrete time HIF is defined as follows:

$$J = \frac{\sum_{k=0}^{N-1} \| n_k - \hat{n}_k \|_{\chi_k}^2}{\| n_0 - \hat{n}_0 \|_{P_0^{-1}}^2 + \sum_{k=0}^{N-1} \left(\| w_k \|_{W_k^{-1}}^2 + \| v_k \|_{V_k^{-1}}^2 \right)}, \tag{9}$$

where N is the number of total time steps, *[Math Processing Error]*$\hat{n}_k$ is the estimated state at time step k, w_k is the state noise, v_k is the observation noise, and n_0 is the initial state, respectively. χ_k, P_k, W_k, and V_k are weighting factors, and $\|\cdot\|$ denotes the vector norm $\left(\text{i.e.,} \| w_k \|_{W_k^{-1}}^2 \ \mathit{impliesw}_k^{\top} W^{-1} w_k \right)$. The way how the weighting factors are determined is that, for instance, if it is known that the second element of $v(k)$ is small, then $V_k(2,2)$ is chosen to be small compared to other elements; otherwise, the weighting factors are determined after preliminary training process. However, the state (n) to be estimated is a scalar in this problem, which results in simplified process of the preliminary tuning process for satisfactory tracking performance.

Direct minimization of J is not tractable; therefore, the performance bound γ is introduced, and it satisfies

$$J < \gamma^{-1}.$$

(10)

Then, J' is defined as

$$J' = -\gamma^{-1} \parallel n_0 - \hat{n}_0 \parallel^2_{P_0^{-1}} + \sum_{k=0}^{N-1} \left[\parallel n_k - \hat{n}_k \parallel^2_{\chi_k} - \gamma^{-1} \left(\parallel w_k \parallel^2_{W_k^{-1}} + \parallel v_k \parallel^2_{V_k^{-1}} \right) \right],$$

(11)

and the problem becomes a matter of solving the following minimax problem:

$$\min_{\hat{n}_k} \left(\max_{w_k, v_k, n_0} J' \right).$$

(12)

The EHIF approach which solves above minimax problem is given by [18]

$$\hat{n}_k = \hat{n}_{k-1} + H_k \left[\xi_k - h(\hat{n}_{k-1}) \right]$$

(13)

if and only if there exists a stabilizing symmetric solution $P_k > 0$ to the following discrete time *Riccati* equation:

$$P_k = P_{k-1} S_k + W_k$$

(14)

where '^' denotes the estimate version, H_k is the H_∞ gain, and given by

$$H_k = P_{k-1} S_k h'(\hat{n}_{k-1}) V_k^{-1}$$

(15)

where $(\hat{n}_{k-1}) = \left. \frac{\partial h(n_k)}{\partial n_k} \right|_{n_k = \hat{n}_{k-1}}$, and

$$S_k = \left\{ I - \gamma \chi_k P_{k-1} + \left[h'(\hat{n}_{k-1}) \right]^2 V_k^{-1} P_{k-1} \right\}^{-1},$$

(16)

where I is the identity matrix or 1 in case of a scalar. Note that *[Math Processing Error]* $h'(\hat{n}_{k-1})$ is employed for an approximated linear form using Taylor expansion due to the non-linearity of the state equation.

Although we need to compute the derivative of the non-linear function $h(n_k)$, it is relatively easy to compute because the state n is a scalar, and consequently, we do not need to compute Jacobian. When the noise attenuation level γ is selected, we have to be very careful, especially it has to satisfy the following condition:

$$\gamma < P_k^{-1} + \left[h'(\hat{n}_{k-1})\right]^2 V_k^{-1} \tag{17}$$

to maintain $P_k > 0$ [19]. We select γ after preliminary training process for optimal performance. The detailed steps of EHIF algorithm for estimating n are summarized in Algorithm 1.

Algorithm 1 EXTENDED H INFINITY FILTERING FOR ESTIMATING *n*

Initialization Initialize the performance bound γ, the estimate $\hat{n}_0$ and weight parameters (χ_k, P_k, W_k, V_k); we select 1 for χ_k.

Recursive update `for` $k = 1, \ldots, N$

(1) Compute H infinity gain
$H_k = P_{k-1} S_k h'(\hat{n}_{k-1}) V_k^{-1}$.

(2) Update the estimate
$\hat{n}_k = \hat{n}_{k-1} + H_k \left[p_k - h'(\hat{n}_{k-1})\right]$.

(3) Update the error variance $P_k = P_{k-1} S_k + W_k$.

In case the system model is linear (could be in some other linear problems), i.e., if (1) and (2) can be expressed in linear forms as follows:

$$n_k = \mathcal{G} n_{k-1} + w_k, \tag{18}$$

$$\xi_k = \mathcal{H} n_k + v_k, \tag{19}$$

where $\mathcal{G}$ and $\mathcal{H}$ are matrixes. Non-extended, original H infinity solution can be obtained by the steps in the following:

1.

H infinity gain $H_k = \mathcal{G}P_{k-1}S_k\mathcal{H}^\top V_k^{-1}$, where $S_k = \left\{I - \gamma\chi_k P_{k-1} + \mathcal{H}^\top V_k^{-1}\mathcal{H}P_{k-1}\right\}^{-1}$.

2.

The estimate $\hat{n}_k = \mathcal{G}\hat{n}_{k-1} + H_k\left[\xi_k - \mathcal{H}\hat{n}_{k-1}\right]$.

3.

Update $P_k = \mathcal{G}P_{k-1}S_k\mathcal{G}^\top + W_k$.

The proposed EHIF approach just adopts Taylor expansion to this solution for approximate linearization of non-linear measurement function. In this problem of estimating n, $\mathcal{G}$ equals 1, and furthermore, $\mathcal{H}$ and *[Math Processing Error]*$\mathcal{H}\hat{n}_{k-1}$ are replaced by *[Math Processing Error]*$h'(\hat{n}_{k-1})$ and *[Math Processing Error]*$h(\hat{n}_{k-1})$, respectively, for the application of extended version of H ∞ filtering approach.

3.3 Computational complexity

We also assess the computational complexity of approaches in big *O* notation in this section. The big *O* notation concerns only major part of computational complexity for comparison purpose. We assume that 'matrix floating point' multiplication demands the highest computational complexity as assumed in [11]. The computational complexity of the approaches are given in the following, when *M* is the number of employed particles, the dimension of the state is *L*, and *G* is the number of GMM mixtures. The EKF requires $O(L^3)$, and the computation of the Kalman gain spans most of calculation time in the algorithm. The EM algorithm requires most time in the GMSPPF algorithm, and its computational complexity is $O(G L^2 M)$. GPF requires the complexity of $O(L^4 M)$, and the calculation of the particle weight takes most time in the algorithm. The computation of S_k demands most time in EHIF filtering, and whole algorithm requires the complexity of $O(L^4)$.

Furthermore, if the state is a scalar, i.e. L=1 with β=2 and G=3, the computational complexity becomes $O(1)$ for the EKF, $O(M)$ for GPF,$O(3M)$ GMSPPF, $O(1)$ for EHIF, respectively. The number of particles are the main factor of the computational complexity for PF approaches. When the number of particles are the same for all PF approaches, GMSPPF requires the highest computational complexity. Although the EKF and EHIF have the same complexity in the big O notation, actually EHIF is much simpler because it does not need the CUSUM algorithm. Due to its low complexity, EHIF has an advantage for implementing in small devices.

4 PERFORMANCE ASSESSMENT

By simulations, we assess and compare the performance of the EKF, GPF, GMSPPF, and EHIF. We consider IEEE 802.11 system's DCF basic access mode where the slot period is 20 μ s, D_{max} is 1024, and the other simulation parameters are the same as selected in [9]. We assume 2,000 observation slots of B, the initial estimate *[Math Processing Error]*$\hat{n}_0 = 5$, and the initial error variance P_0=10 for all filtering methods. The CUSUM algorithm is adopted in the EKF, GPF, and GMSPPF. Three hundred particles are employed for GPF and GMSPPF. The parameters for GMSPPF and EHIF are as follows: G=3, I=1, J=2, γ=0.001, χ_k=1, W_k=2, and V_k=0.0001. The simulations are performed using MATLAB 7.11.0. Some *ReBEL Toolkit* functions (e.g., GMM) are adopted for GMSPPF as employed in [16]. The performance is assessed via 'MSE of estimated n' and the 'saturation throughput' which has been defined in Section 2.

Figure 2 shows how D_{min} affects the saturation throughput performance. In the middle of simulation, n varies from 10 to 25. At that time, $D_{min}$is adjusted based on two numbers of terminals according to (4), i.e., the true n and the estimate *[Math Processing Error]*$\hat{n}^{EKF}$ by the EKF, respectively. Whereas the throughput is degraded as n increases with non-adjusted D_{min}, the throughput degradation is not observed when D_{min} is adjusted based on true n and *[Math Processing Error]*$\hat{n}^{EKF}$.

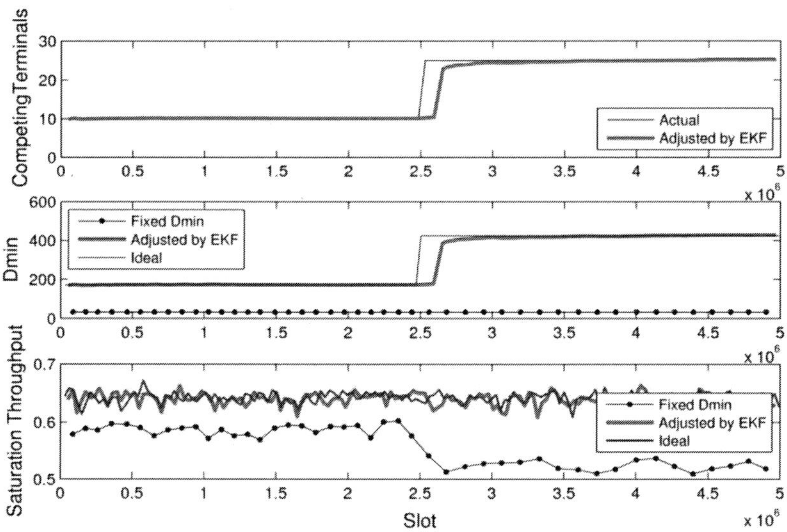

Figure 2. Saturation throughput performance depending on D_{min}.

In the saturated network condition, all terminals are assumed to always have a packet waiting for transmission in their buffers. We assume that n varies with time as follows: (5,10,25, and 15 terminals) and simulation time duration is 350 s. This scenario is similar to that employed in [9]. We run 200 simulations for both MSE and throughput performance assessment. In this scenario, all terminals estimate n and reflect it for adjusting D_{min} to maximize the throughput. The result shows that EHIF outperforms all the other approaches in both MSE and saturation throughput as shown in Table 1. The result shows similar throughput performance for all approaches during the period of non-variation of $n$whereas EHIF shows the best throughput performance during the period of variation of n. In Table 1, we highlighted the throughput performance of EHIF during the period of the largest variate of n (i.e., from 10 to 25 at 150 s). This result is from the fact that the approaches that employ the state variation detector based on CUSUM algorithm has considerable latency to make the decision for adjusting the variance in their filtering. On the other hand, EHIF approach does not have any latency during any variation of n because any decision is not needed to be made in its filtering process. This is how EHIF avoids the latency and makes the prompt tracking possible for consequent enhanced performance.

Table 1. Results of saturated network condition

	Mean squared error		Saturation throughput (150 to 160 s)	
	Mean	Variance	Mean	Variance (10^{-5})
EKF	1.492829	0.055072	0.634940	4.891636
GPF	1.105716	0.034910	0.635292	3.649367
GMSPPF	1.079393	0.032417	0.635308	3.179673
EHIF	0.894879	0.019242	0.637274	1.853646

In the non-saturated network condition, all terminals send packets randomly with random lengths. The scenario is close to the real internet network condition where users' random requests generate random data traffic. Therefore, prompt tracking and latency avoidance is more critically demanded in this scenario than the case of the non-saturated network condition. We assume that the data traffic length and the transmission period follow the exponential distribution with the rates of 0.11/s and 0.7/s, respectively. To describe large and frequent variations of *n*, e.g., as in subways or hotspot areas, we change *n* every 20 s in the pattern of (15,25,15,30,and 20 terminals). Figure 3 shows an example of tracking *n* by the EKF and EHIF where we can notice that the tracking performance of EHIF is better than that of the EKF. The EKF shows latency clearly in tracking *n* that is caused by the employment of the state variation detector in the algorithm. On the other hand, EHIF shows prompt tracking in response to the variation of the state *n* without any latency. We evaluate MSE of estimates and saturation throughput by 200 runs with this scenario. Figures 4 to 5 and Table 2 show that EHIF outperforms all the other filtering approaches in the non-saturated network condition. EHIF shows the best performance, particularly during the periods of the large variate of *n* due to its prompt tracking performance. Figure 5 shows the throughput performance of the methods, and the performance comparison is noticeable clearly only during the periods of the large variate of *n*. During the periods of non-varying *n*, significant difference of the throughput performance is not observed unless the difference between the optimal $D_{\min}$ and applied one is greater than a certainly impactive level. Nonetheless, prompt tracking ability is a crucial factor for enhanced throughput performance, particularly for the periods of dynamic variation of n.

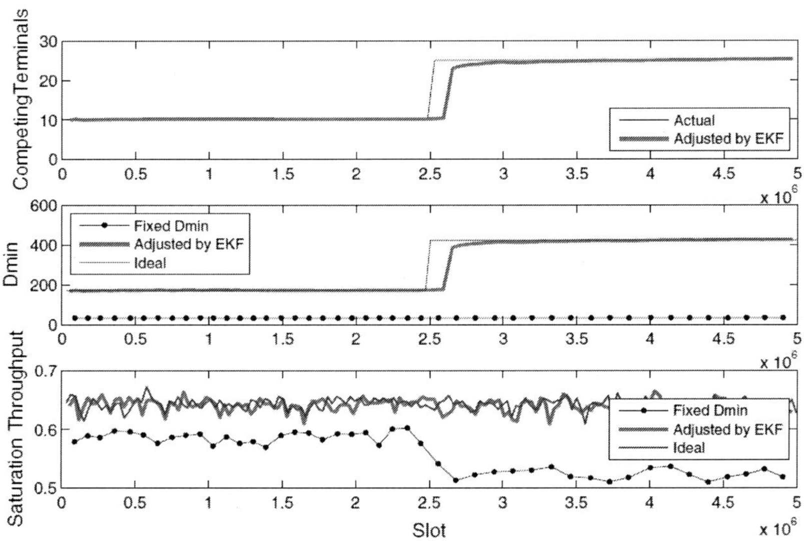

Figure 2. Saturation throughput performance depending on *D*min.

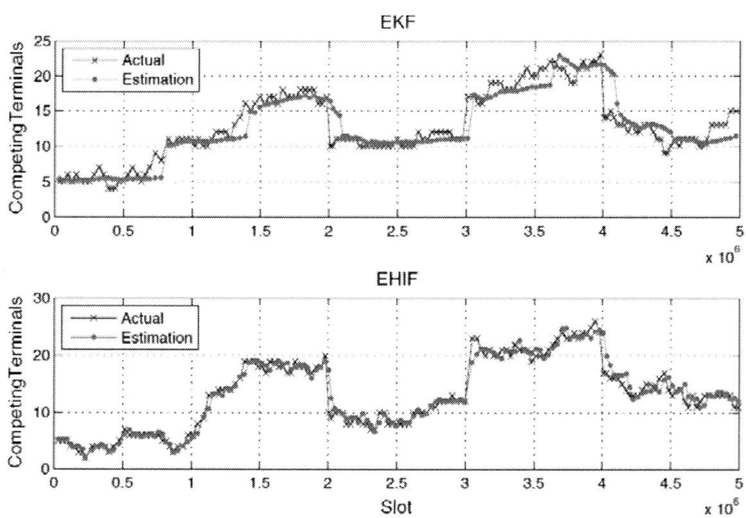

Figure 3. Example of tracking *n* by the EKF and EHIF in non-saturated network condition.

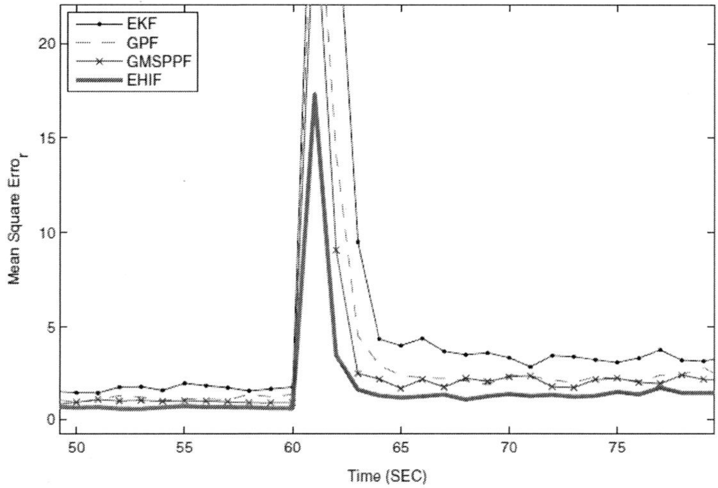

Figure 4. Mean squared error in non-saturated network condition.

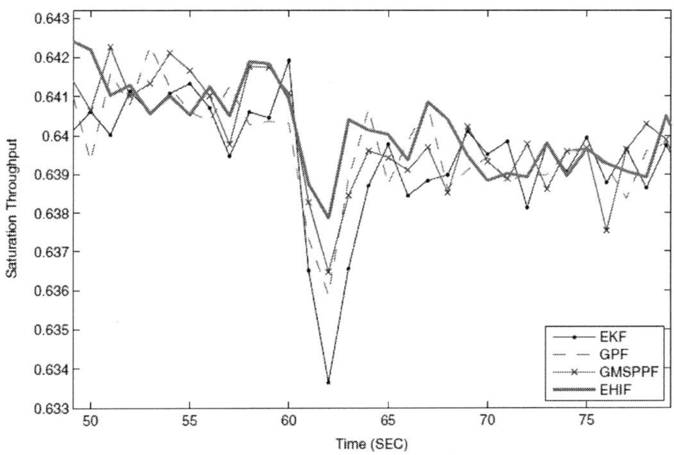

Figure 5. Saturation throughput in non-saturated network condition.

Table 2. Results of non-saturated network condition

	Mean squared error		Saturation throughput	
	Mean	Variance	Mean	Variance (10^{-7})
EKF	4.363303	0.792361	0.639690	8.408113
GPF	2.818697	0.391450	0.640170	8.687158
GMSPPF	2.432454	0.292157	0.640205	7.770071
EHIF	1.528985	0.110101	0.640418	7.325334

5 CONCLUSIONS

In this paper, we proposed the EHIF approach for estimating the number of competing terminals to enhance the throughput performance of the IEEE 802.11 system. We showed that EHIF outperformed previously proposed approaches (i.e., EKF and GMSPPF) in both saturated and non-saturated network conditions. We focused on the 'non-saturated condition' more because the problem is barely investigated for the non-saturated condition in the literature. Furthermore, the non-saturated condition is closer to real scenarios because the state n varies more dynamically in the condition than it varies in the saturated condition. In the application of EHIF, the knowledge of noise statistics is not needed; therefore, additional CUSUM algorithm is not needed that incurs the considerable latency in tracking n, which results in significantly degraded tracking performance. Consequently, we obtained improved saturation throughput performance by employing EHIF based on prompt tracking of n. Furthermore, EHIF can be easily implemented in small devices due to its low computational complexity.

ABBREVIATIONS

CSMA/CA: Carrier sense multiple access with the collision avoidance

CTS: Clear-to-send

CUSUM: Cumulative summary

DCF: Distributed coordination function

DIFS: Distributed inter-frame space

EHIF: Extended H infinity filter

EKF: Extended Kalman filter

EM: Expectation maximization

GMM: Gaussian mixture model

GMSPPF: Gaussian mixture sigma point particle filter

GPF: Gaussian particle filter

HIF: H infinity filter

MAC: Medium access control

MSE: Mean squared error

PF: Particle filter(ing)

PHY: Physical layer

RTS: Request-to-send

SIFS: Short inter-frame space

UKF: Unscented Kalman filter

WLAN: Wireless local area network.

ACKNOWLEDGEMENTS

This research was supported by the Basic Science Research Program through the National Research Foundation of Korea funded by the Ministry of Education (NRF-2011-0009255), the MSIP (Ministry of Science, ICT & Future Planning), Korea in the ICT R &D Program 2013, and the MSIP,

Korea, under the IT Consilience Creative Program (NIPA-2013-H0203-13-1001) supervised by the National IT Industry Promotion Agency.

REFERENCES

1. IEEE: *Wireless LAN Medium Access Control (MAC) and Physical Layer (PHY) Specifications*. Piscataway: IEEE; 1999.
2. Romaszko S, Blondia C: Cross layer PHY-MAC protocol for wireless static and mobile ad hoc networks. *EURASIP J. Adv Signal Processing*2009, 2009: 278041. 10.1155/2009/278041
3. Hewer T, Nekkovee M, Coveney P: Reducing congestion in obstructed highways with traffic data dissemination using ad hoc vehicular.*EURASIP J. Adv. Signal Process* 2010, 2010: 169503. 10.1155/2010/169503
4. Bianchi G, Fratta L, Oliveri M: Performance evaluation and enhancement of the CSMA/CA MAC protocol for 802.11 wireless LANs. *Proc, PIMRC 1996* 1996, 2: 392-396.
5. Hwang G, Cho D: Performance analysis on coexistence of EDCA and legacy DCF stations in IEEE 802.11 wireless LANs. *IEEE Trans. Wireless Commun* 2006, 5(12):3355-3359.
6. Chang S, Wu H: Novel adaptive DCF protocol using the computationally-efficient optimization with the feedback network information for wireless local-area networks. *IEEE Trans. Wireless Commun* 2009, 8(6):2827-2830.
7. Bianchi G: Performance analysis of the IEEE 802.11 distributed coordination function. *IEEE J. Select. Areas Telecommun., Wireless Series*2000, 18(3):535-547.
8. Deng D, Ke C, Chen H, Huang Y: Contention window optimization for IEEE 802.11 DCF access control. *IEEE Trans. Wireless Commun*2008, 7(12):5129-5135.
9. Bianchi G, Tinnirello I: Kalman filter estimation of the number of competing terminals in an IEEE 802.11 network. *Proc. Infocom 2003* 2003, 2: 844-852.
10. Vercauteren T, Toledo AL, Wang X: Batch and sequential Bayesian, estimators of the number of active terminals in an IEEE 802.11 network. *IEEE Trans. Signal Process* 2006, 55(2):437-450.
11. Kim J, Serpedin E, Shin D: Improved particle filtering-based estimation of the number of competing stations in IEEE 802.11 networks. *IEEE Signal Process. Lett* 2008, 15: 87-90.
12. Arulampalam M, Maskell S, Gordon N, Clapp T: A tutorial on particle filters for online nonlinear/non-Gaussian Bayesian

tracking. *IEEE Trans. Signal Processing* 2002, 50(2):174-188. 10.1109/78.978374
13. El-Halym HAA, Mahmoud I, Habib S: Proposed hardware architectures of particle filter for object tracking. *EURASIP J. Adv. Signal Process*2012, 2012: 17. 10.1186/1687-6180-2012-17
14. Jishy K, Lehmann F: A Bayesian track-before-detect procedure for passive radars. *EURASIP J. Adv. Signal Process* 2013, 2013: 45. 10.1186/1687-6180-2013-45
15. Bianchi G, Tinnirello I: Interference estimation in IEEE 802.11 networks. *IEEE Control Syst. Mag* 2010, 30(2):30-43.
16. van der Merwe R, Wan E: Gaussian mixture sigma-point particle filters for sequential probabilistic inference in dynamic state-space models. *Proc. Int. Conf. Acoustics, Speech, and Signal, Process. (ICASSP)* 2003, 6: 701-704.
17. Kotecha JH, Djuric PM: Gaussian particle filtering. *IEEE Trans. Signal Process* 2003, 51(10):2592-2601. 10.1109/TSP.2003.816758
18. Shen X, Deng L: Game theory approach to discrete H_∞ filter design. *IEEE Trans. Signal Process* 1997, 45(4):1092-1095. 10.1109/78.564201
19. Jing Y, Yin F, Chen Z: An H_∞ filter based approach to Coombat inter-carrier interference for OFDM systems. *IEEE Commun. Lett* 2008, 12(6):453-455.

2
CHAPTER

DEVELOPMENT OF LONG-RANGE AND HIGH-SPEED WIRELESS LAN FOR THE TRANSMISSION OF TELEMEDICINE FROM DISASTER AREAS

Masayuki Nakamura,[1] Shoshin Kubota,[1] Hideaki Takagi,[1] Kiyoshi Einaga,[2] Masashi Yokoyama,[3] KatsutoMochizuki,[4] Masaomi Takizawa,[5] and Sumio Murase[5]

ABSTRACT

A computer network is indispensable for realizing the use of telemedicine. Recently, experiments to provide telemedicine to residents in remote places over a broadband Internet access have been reported. However, if a disaster were to occur with devastation over a wide mountainous area, and telephones and Internet access were to become unavailable, the provision of telemedicine for injured residents in this area becomes difficult. To solve this problem, we have developed 2.4 GHz wireless LAN units with the longest coverage in Japan to date, of 30 km plus at 54 Mbps which complies with the IEEE802.11 g standard and the Japanese radio regulations to re-establish communications temporally between disaster devastated areas and hospitals, and so on. We tested them in the disaster prevention drill with the regional fire bureau and concluded that wireless LAN units we developed can transfer high-quality video images and sound good enough for use in telemedicine.

1. INTRODUCTION

Providing prompt and efficient post disaster measures, the opportunity to share and utilize live video images of a devastated area has become very important for disaster-related organizations.

Emergency services, such as fire, search and rescue, hospitals, and international help organizations will also need to provide much needed assistance to help victims in a catastrophe. They will also need to use an affordable and reliable communication system to help people in remote and devastated areas.

In the aftermath of a disaster, breakdowns in communication networks usually occur. Examples would be the affected areas in the 2004 tsunami in the Indian Ocean, and the 2005 Hurricane Katrina incident in the United States. Both incidents highlight the need for a communication system that is easy to set up, operate, and that is also inexpensive.

After these disasters, it became apparent that the failure in the communication networks greatly contributed to the high number of casualties, injuries, and sickness in the affected areas.

The importance of providing telemedicine in Japan is starting to be realized, as this country is fast becoming an aged society. By 2050, it is predicted that more than 40% of Japan's population will be over the retirement age of 60.

Providing telemedicine with high-quality video conferencing using broadband networks, mainly used for Internet access, has been studied, as fiber-optic networks have spread and become available in Japan [1].

To provide telemedicine to areas where no broadband network service is provided or where any type of telecommunication becomes unavailable, the only infrastructure that is expected to be accessible is through a satellite network.

One reason is because cellular phones are eliminated from service for two reasons. The first is that towers for base stations of the cellular phone systems have collapsed and electric power outage occurs. Or the system is overloaded with nonemergency telephone traffic from outside the affected areas and thus is shutdown.

The use of satellite communication has been proposed and studied to be used for telemedicine and in disasters [2–4].

With the advance of satellite communication equipment, satellite IP network services using very small aperture terminal (VSAT) have become popular in mainly rural areas to access the Internet. And VSAT has been studied for providing telemedicine [5, 6].

To establish temporal network connections especially after a disaster, a hybrid network communication system using VSAT and wireless fidelity (WiFi, IEEE802.11b/g wireless LAN) has been proposed and evaluated [7, 8]. This type of system has already been field-tested and put on the market.

However, satellite network connection services by VSAT have three major problems; they are as follows.

It provides asymmetrical transmission rates. Even in advanced services using a 1.2 m diameter dish, uplink speed is less than about 2 Mbps, even though downlink speed is up to 60 Mbps.

In providing telemedicine, uplink speeds are an important feature since high-quality still or live video images of patients should be transferred to hospitals using uplink communication. These images are necessary to enable doctors to diagnose victims' medical conditions.

The second problem is that VSAT equipment is heavy. When a large-scale natural disaster occurs, it may be impossible to take VSAT equipment to some devastated areas, since roads are greatly damaged and all traffic is shut down.

The third is that there is at least 1 country where VSAT is not licensed to be used for telemedicine [9], and there may be more.

Beside applications of satellite communications in disasters, use of WiFi and Worldwide Interoperability for Microwave Access (WiMAX) has been proposed and studied [10, 11].

It has been assumed that WiMAX can provide broadband wireless access (BWA) up to 50 km between fixed stations and 5–15 km for mobile stations. And it can provide high-transmission rates of up to 70 Mbps [12, 13]. Of course, there is a tradeoff between the distance and transmission rates.

WiMAX is now in its early stage of deployment and is not widely implemented. Most evaluations have been conducted on an experimental basis [14]. As WiMAX is a metropolitan area network (MAN) technology, its deployment has started in urban areas thus deployment in rural areas has a high possibility of being delayed.

In most countries, WiMAX uses licensed spectrum in 2.5 GHz band and/or 3.5 GHz band. WiMAX infrastructures will be very similar to cellular phones' infrastructures. WiMAX-base stations are to be mounted on towers and rooftops of buildings. All base stations are connected to each other with wired high-speed backhaul networks. As a result, large-scale disasters cause significant damages to WiMAX infrastructure and it would be expected that big troubles like cellular phone systems are sustained.

On the other hand, WiFi or IEEE802.11b/g wireless LAN is said to have shorter coverage of up to some 100 m and it can provide transmission rates of up to 54 Mbps.

Using WiFi units in disaster areas, establishing makeshift broadband networks has also been studied. Researches on setting up mesh networks or ad hoc networks with WiFi units in areas struck by disaster and evaluation of its networks have been conducted [15–17]. WiFi has advantages over satellite and WiMAX in the following areas.

Using commercial off-the-shelf WiFi units, high-speed network almost equivalent to WiMAX can be established at lower costs. WiFi units are available all over the world with the same unlicensed spectrum. Therefore, fire, hospitals, and search and rescue teams from overseas that usually come with limited communication capability can communicate to each other with a high-speed transmission capability.

In Japan, the central government, local governments, fire offices, and hospitals use different licensed spectrums and thus are not permitted to communicate to each other. WiFi can remove this restriction.

A shortcoming of WiFi is the coverage. Most researches were conducted in less than 300 m coverage [18, 19].

To solve this problem and enable WiFi to be used in a disaster, we developed long-range and high-speed wireless LAN units to establish makeshift but high-speed wireless LAN networks. These networks would establish communication between devastated areas and local authorities.

The networks are easy to be built and anyone can use without any radio-related licenses. Our networks would also be practical for use in other countries with limited resources and manpower to establish temporary emergency communications.

To demonstrate the capability of this unit, we helped organizing an earthquake drill conducted with a regional fire bureau in Japan. We established a communication network using these units which gave a video and audio link between two points at a distance of over 40 km.

This link was between a village located in a valley surrounded by tall mountains and a hospital in an urban area located in a different valley.

From this emergency drill, our experiments succeeded in being the first of its kind to transmit video and audio in this way. We also establish a record for the greatest distance that information was transmitted in Japan.

We believe this system can be used over a greater distance if it were operational in nonmountainous terrain.

This paper is organized as follows. Section 2 describes the deinformation velopment of long-range and high-speed wireless LAN units. Section 3 describes an experiment on a telemedicine network in an emergency with wireless LAN units. Section 4 describes experiment results and Section 5 presents discussions. Finally, Section 6 draws conclusions.

2. DEVELOPMENT OF LONG-RANGE AND HIGH-SPEED WIRELESS LAN UNITS

Most of the information provided in this paper will center on Japan, because most of our research and experiments were undertaken in that country.

First of all, we introduce the long-range and high-speed wireless LAN unit we developed to establish a temporary emergency link. This type of communication system enables information transfer between mountain areas and the disaster-related organizations such as hospitals and local authorities.

Currently in Japan, under the domestic radio regulations, wireless LAN is the only radio equipment with a high-transmission rate permitted to be used without any radio-related licenses for the operation.

This allows local residents, firefighters, paramedics, doctors, nurses, and any foreign relief agencies to operate this setup without getting any radio-related licenses.

Summary of the benefits in using wireless LAN for telemedicine in emergency situations are given as follows:

1. no radio-related licenses are required to operate in Japan or other countries; anyone, any organization can set up networks, operate and communicate to each other;

2. very low cost and setting up time are anticipated to establish a wireless LAN network, even in developing countries; compare this

to some of the equipment that may be needed or used by telecommunication companies in providing high-speed (broadband) networks;

3. smooth and efficient connectivity with TCP/IP networks;
4. available use of all software (video-conferencing software, etc.) and equipment (video cameras and hookup wires, e.g.) for use with personal computers.

So far, we developed IEEE802.11b-complied-wireless LAN units. These units have a capability of coverage of about 50 km at the transmission rate of 11 Mbps.

By using these units, we have succeeded in giving network communication links between mountain huts and hospitals at the foot of these mountain areas to provide telemedicine to mountain climbers for more than 3 years [20–22].

These mountain huts are located in the Japan Alps at an altitude of 3000 m or thereabout, 250 km west of Tokyo. The Japan Alps is an area where no broadband network service was available, although more than 250 000 climbers from all over the world visit this area in summer season alone.

As high definition, TV broadcasting recently has become popular in Japan, demands for video transmission with higher picture quality and multiple video images for use in telemedicine have increased. The IEEE802.11b-complied-wireless LAN does not have enough capability to transfer higher-quality video.

To meet these demands, IEEE802.11g-complied-wireless LAN is promising, as this type of wireless LAN has a faster transmission rate than IEEE802.11b-complied units.

However, since commercially available wireless LAN units for 2.4 GHz band in Japan are designed to be used indoors. Their coverage is usually for a very short range of about 3 km when trying to operate them with a high-gain antenna at a speed of 54 Mbps.

As a result, it is almost impossible to link between mountain areas and urban areas where the local governments and hospitals are usually located using this type of arrangement.

Long-range wireless LAN units we have developed, which comply with the IEEE802.11g standard and Japanese radio regulations, can cover a distance of up to 30 km at a transmission rate of 54 Mbps.

2.1. Extension of Wireless LAN Coverage

Under current Japanese radio regulations for 2.4 GHz band wireless LAN (IEEE802.11b/g), a radiation power of 10 mW/MHz with a 12.14 dBi antenna is allowed under the condition that the half-power beamwidth of an antenna is less than 30 degrees. This indicates that the maximum equivalent isotropically radiated power (EIRP) per MHz should be 22.14 dBm/MHz. Figure 1(a) shows this maximum EIRP.

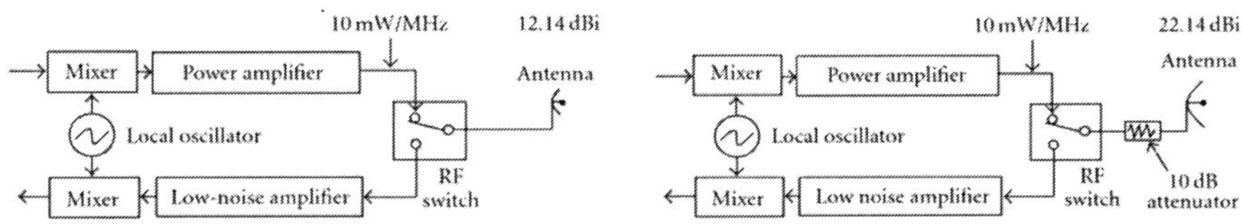

Figure 1. OUTPUT POWER AND ANTENNA GAIN: (A) CURRENT JAPANESE REGULATION FOR EIRP, (B) AN ATTENUATOR USED WITH A HIGHER GAIN ANTENNA.

To use an antenna with a higher gain, we have to reduce to the maximum EIRP to 22.14 dBm/MHz. An attenuator is inserted between the antenna and RF switch in almost all cases to meet this regulation.

As shown in Figure 1(b), to use an antenna with a gain of 22.14 dBi, a 10 dB attenuator is inserted. As an antenna with higher gain has a narrower half beamwidth, this method has the benefit of reducing interferences from nearby wireless LAN units. However, it sacrifices the reception sensitivity; as a result, this method gives a slightly longer coverage.

Our method to extend the coverage does not use this attenuator. Thus, this method can transfer signals received by an antenna to a low-noise amplifier with minimum losses. As an attenuator is not used, the output power should be reduced to 1 mW/MHz as shown in Figure 2.

However, this method has the following problems.

1. Need to control the power amplifier to output very low-level signals to keep stable.
2. Need to reduce the leakage radiation of a carrier signal from an antenna, which is produced by a local oscillator and leaks through the low-noise amplifier to the antenna.

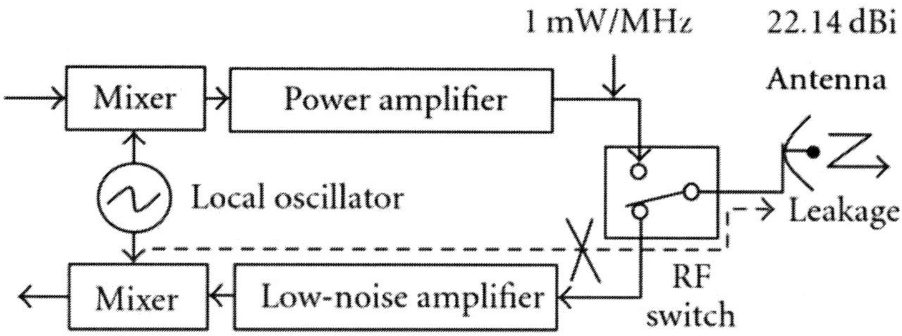

Figure 2. OUR COVERAGE-EXTENSION METHOD.

We succeeded in solving these problems. And the use of a higher-gain antenna was achieved.

2.2. Developed Wireless LAN Unit

In our development, we succeeded to use 26.5 dBi antenna to receive weaker signals under this regulation.

Key features of our newly developed wireless LAN unit are as follows:

1. complies with the Japanese radio regulation for 2.4 GHz wireless LAN and IEEE802.11b/g standards;
2. maximum transmission rate: 54 Mbps (48, 36, 24, 18, 11, 9, 6, 5.5, 2, 1 Mbps available depending on the distance and propagation conditions);
3. frequency range: 2400 MHz–2483.5 MHz;
4. coverage: up to 30 km at 54 Mbps with 26.5 dBi antenna (further with lower transmission rates);
5. interface for network: ethernet/power on ethernet;
6. low power consumption: about 10 W;
7. 24 V battery operational.

Figure 3 shows the wireless LAN unit. Its size is about 20 cm × 20 cm and it weighs about 1.6 kg.

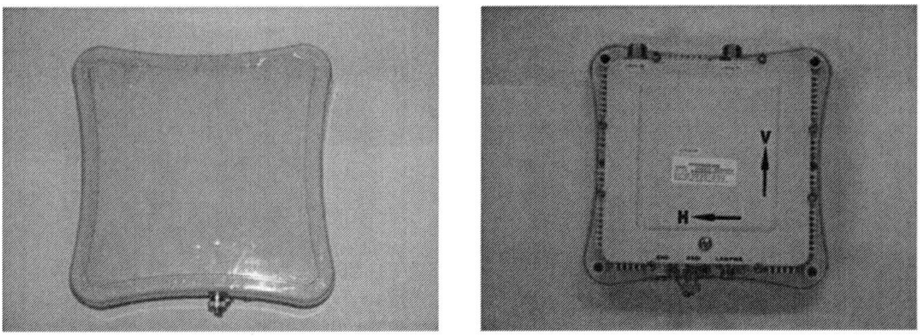

Figure 3. A WIRELESS LAN UNIT DEVELOPED. Front viewback view

Power supply is either 24 VDC or power on ethernet (POE) with 100Base-T and consumes about 10 W on average. This unit can also be operated with small batteries (car batteries) charged by solar cells because of its low power consumption.

A portable parabolic antenna with a gain of 24 dBi we used in this experiment with this wireless LAN is about 100 cm × 60 cm and weighs only 3.5 kg.

As a result, it is very easy to take these wireless LAN units, antennas, batteries, and so on, to areas devastated by the disaster.

Figure 4 shows a typical installation of this wireless LAN unit and this 24 dBi parabolic antenna to a steal pipe.

These units can be transported by people on foot to establish a temporary wireless communication network capable of a high-speed transmission rate.

Because these units are so easy to transport, set up, and operate, they can be installed and operated in virtually any terrain by almost any person, with minimum instruction.

Figure 4. AN INSTALLATION EXAMPLE OF A PARABOLIC ANTENNA AND A WIRELESS LAN UNIT TO A STEAL PIPE.

Because these units are so manageable, valuable transportation assets are not necessarily needed and can be used where they are more in demand for search and rescue or relief operations.

As mentioned in the introduction, this unit is the first of its kind and currently has the longest coverage in Japan to date.

3. EXPERIMENTS ON A TELEMEDICINE NETWORK IN AN EMERGENCY WITH WIRELESS LAN UNITS

One of our experiments was conducted, as a part of an earthquake drill, and it was organized by a regional fire bureau and a village office in Japan. This is reflecting several large-scale earthquakes that have recently occurred in Japan, which killed a lot of people.

Nagano Prefecture General Industrial Technology Center (hereafter referred to as NPGITC) and Shinshu University Hospital joined the drill to demonstrate the capability of this wireless LAN.

This earthquake drill was planed on the supposition that a strong and large-scale earthquake occurred in a mountainous area and all possible communication networks to this area become unavailable.

The purpose of this drill was to take all possible measures in mitigating the damage of the earthquake and to reinforce the rescue measures for victims.

It especially focuses on helping paramedics give emergency treatment to victims in remote places using a makeshift but high-speed network.

Paramedics guided by doctors, located in hospitals, watching live video images of victims being sent over the wireless LAN network from the devastated areas, administer aid to patients.

In the case of Japan, paramedics are limited to the degree that they can treat and give medical help to patients. In some scenarios, paramedics can only perform treatments under a doctor's supervision. Our network allows emergency staff in the field to extend range in giving treatments under a doctor's control.

We believe that there may be other countries or regions where emergency workers are limited in the amount of aid that they can deliver without the permission or guidance of a medical doctor.

Our equipment may be useful in similar situations allowing paramedics to work with doctors to save more lives.

3.1. Long-Range Wireless LAN Network Established

Participants and locations for the aforementioned earthquake drill are as follows.

1. Main drill site:
   (a) elderly care center "Mizuki" (in Omi village, Nagano Prefecture).

2. Organizations:
   (a) Matsumoto Regional Fire Bureau (in Matsumoto City, Nagano).
   (b) Shinshu University Hospital (in Matsumoto City, Nagano).

(c) Nagano Prefecture General Industrial Technology Center (NPGITC) (in Matsumoto City, Nagano).

The wireless networks made up of these wireless LAN units are shown in Figure 5.

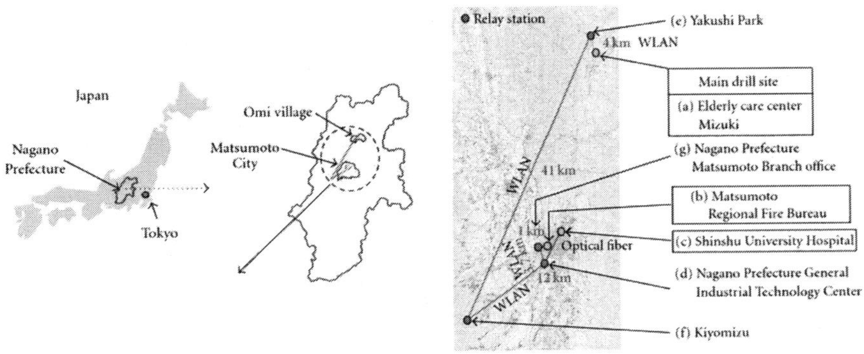

Figure 5. Temporary wireless LAN network.

The wireless LAN in 2.4 GHz band should be operated in line of sight. The area between the main drill site and Shinshu University Hospital in our experiment is not in line of sight. So we placed 4 relay stations between Matsumoto Regional Fire Bureau and the main drill site as follows (also see Figure 5):

(e) Yakushi Park in Omi village,

(f) Kiyomizu located in Yamagata village, Nagano,

(d) Nagano Prefecture General Industrial Technology Center (NPGITC),

(g) Nagano prefecture Matsumoto branch office in Matsumoto city, Nagano.

The following are the distances between each place where wireless LAN units are located:

1. Main drill site (a) and Yakushi Park (e): 4 km;
2. Yakushi Park (e) and Kiyomizu (f): 41 km;
3. Kiyomizu (f) and NPGITC (d): 12 km;

4. NPGITC (d) and Nagano prefecture Matsumoto branch office (g): 3.7 km,

5. Nagano prefecture Matsumoto branch office (g); and Matsumoto Regional Fire Bureau (b): 1 km.

Forty one kilometers is the longest distance covered in this experiment, between Yakusi Park and Kiyomizu.

At each relay station, two wireless LAN units are connected to each other by its Ethernet interface via a switching hub as shown in Figure 6. Connections are simple and very easy to be completed by just about anyone.

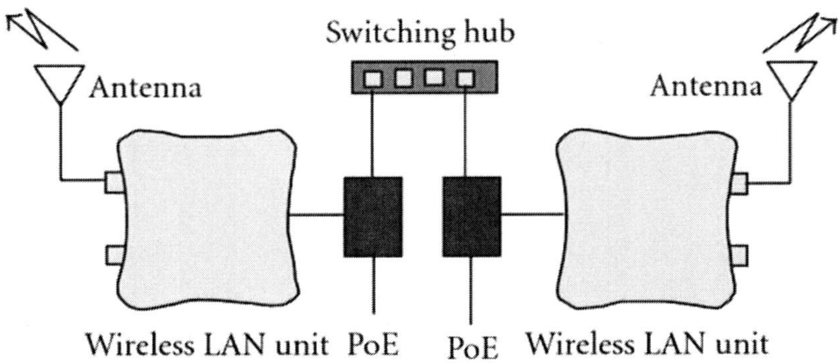

Figure 6. A RELAYING METHOD.

The entire established network when seen from the viewpoint of the wireless network connection is shown in Figure 7.

Between NPGITC (d) and Shinshu University Hospital (c) alone, a fiber-optic network with a 100 Mbps transmission rate was used. This optical fiber was provided by a local CATV company.

3.2. High-Quality Live Video Image Transmission from the Main Drill Site

Figure 8 shows a high-quality live video image transmission network from the main drill site to Shinshu University Hospital and the Matsumoto Regional Fire Bureau.

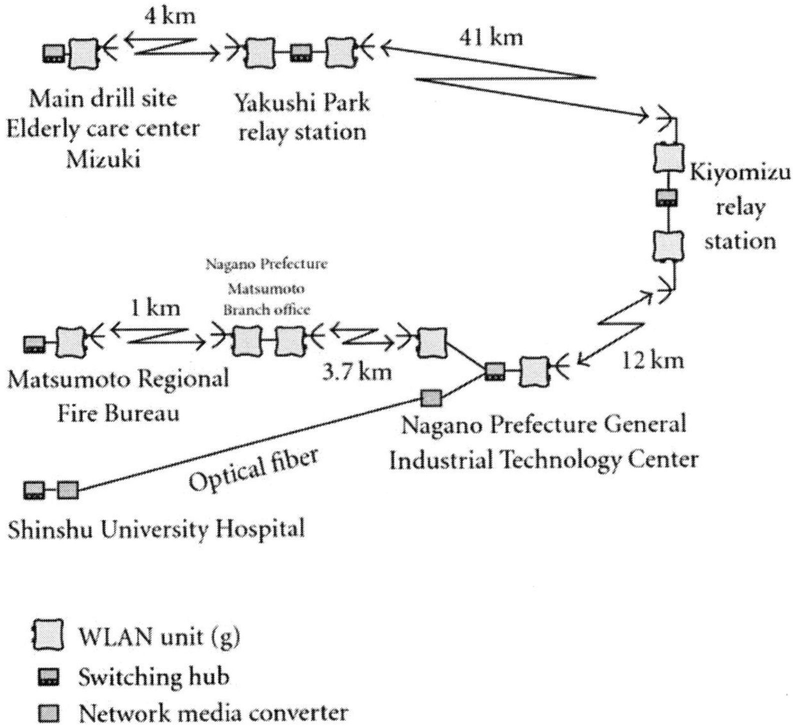

Figure 7. WIRELESS LAN NETWORK ESTABLISHED.

This network consists of three systems. The first is the main wireless LAN network aforementioned in Figure 7 to provide network connections. The second is a portable live video transmission system for use in the main drill site. The third is a compact helicopter live video transmission system to send video images of the drill site.

The portable live video transmission system is to send detailed information of the devastated areas and patients' medical conditions. The compact helicopter live video transmission system is for giving the overview of the devastated areas for doctors and the fire office.

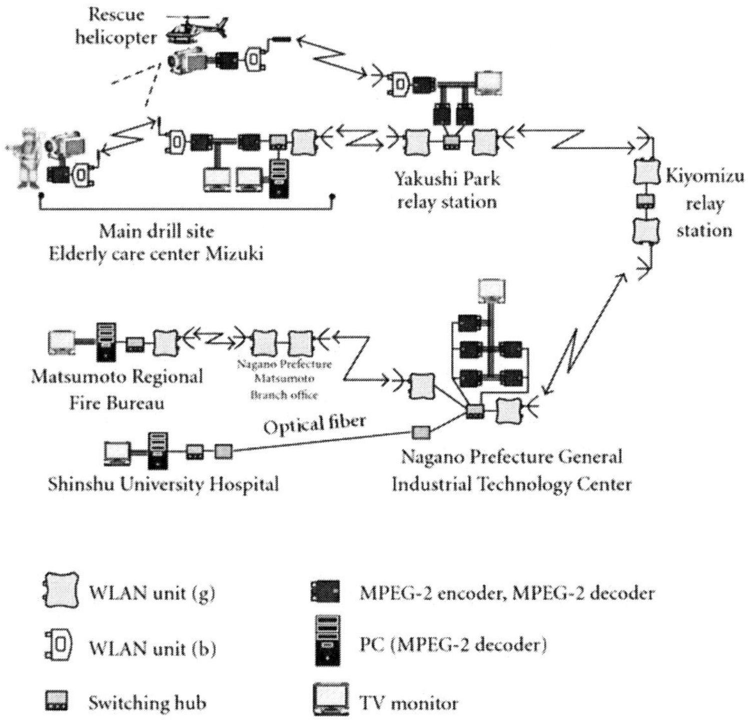

Figure 8. LIVE VIDEO IMAGE TRANSMISSION NETWORK.

We developed this portable live video transmission system using IEEE802.11b wireless LAN units aforementioned.

Firefighters are able to send high-quality live video images and sound from anywhere in the main drill site area by walking with a video camera.

This system has a capability of transmitting video images and stereo sound with quality almost as high as TV broadcasting.

We also developed this compact helicopter live video transmission system using the same IEEE802.11b wireless LAN units.

This system also can send high-quality live video images and sound taken from inside a helicopter to the ground.

Figure 9 is a schematic of this live video transmission network focusing on encode and decode of video and audio signals.

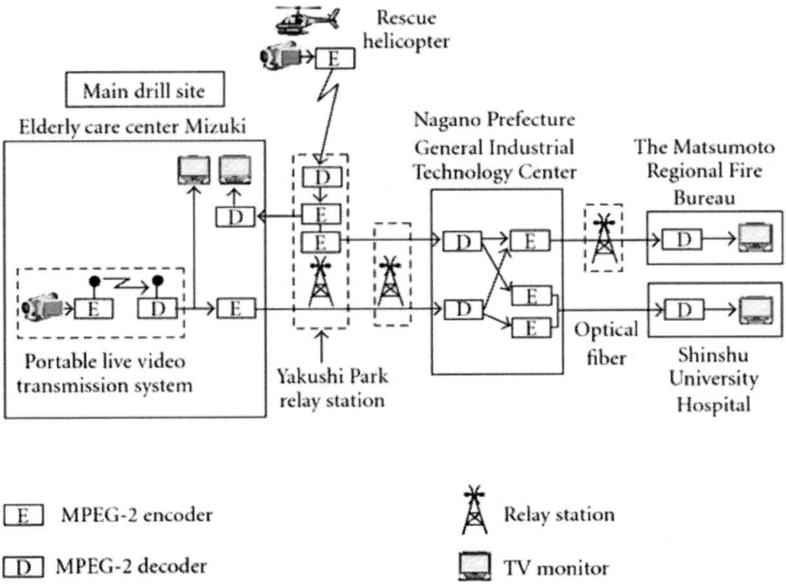

Figure 9. BLOCK DIAGRAM OF VIDEO ENCODE AND DECODE.

The live video images and sounds sent from the portable video-transmission system are received at the base station in the main drill site, and are then sent to the Yakushi Park relay station. Yakushi Park relay station then forwards these packets to the Kiyomizu relay station.

In Kiyomizu relay station, the received video image and sound packets are forwarded to the NPGITC. At NPGITC, these image and sound packets are decoded to video and sound signals and these signals are encoded and resent to the Matsumoto Regional Fire Bureau and Shinshu University Hospital at the same time.

Meanwhile, live video images and sounds sent from the helicopter are received at Yakushi Park relay station. These image and sound packets are decoded to video signal and sound signals, and displayed. Then these signals are encoded again to be sent to the main drill site, Matsumoto Regional Fire Bureau, and Shinshu University Hospital.

Two different encoders are used at Yakushi Park relay station to send live video images to two different directions, the main drill site, and Matsumoto Regional Fire Bureau or Shinshu University Hospital.

At the main drill site, live video images sent from a firefighter and a helicopter can be seen at the same time.

At Nagano Prefecture General Industrial Technology Center (NPGITC), live video images and sounds sent from the main drill site and the helicopter are decoded to video and sound signals. These two types of signals are encoded by three encoders and sent to Matsumoto Regional Fire Bureau and Shinshu University Hospital.

Two encoders are used for Shinshu University Hospital in order that live video images sent from the main drill site and the helicopter can be seen at the same time.

The third encoder is for Matsumoto Regional Fire Bureau, which has 2 video and stereo audio interfaces.

At Matsumoto Regional Fire Bureau, either live video images sent from the main drill site or the helicopter can be seen. This selection is done by changing input interface of the encoder at NPGITC remotely from Matsumoto Regional Fire Bureau.

As shown in Figure 9, to transmit live video and audio signals from the main drill site to Shinshu University Hospital, three pairs of encoders and decoders were used.

All encoders and decoders used in each system encode and decode MPEG-2 video and sound streams.

Figure 10(a) shows the encoder and decoder units we mainly used in this drill.

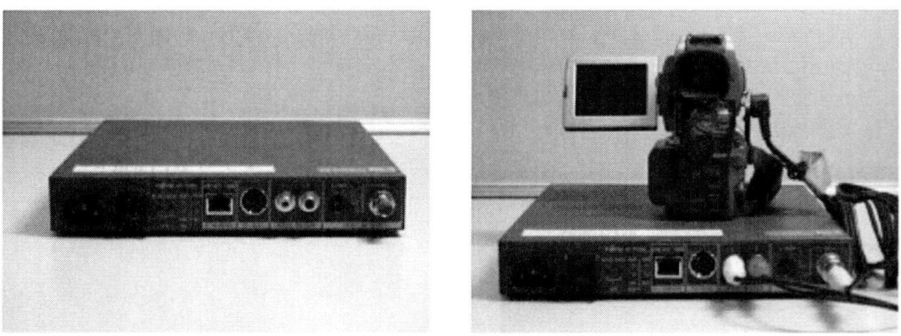

Figure 10. MPEG-2 ENCODER AND DECODER: (A) ENCODER, DECODER UNIT, (B) CAMCORDER WITH ENCODER.

These encoders and decoders have the same shape. By changing the software to be installed, these units can be either an encoder or a decoder. This unit has video and sound interface as shown in Figure 10(b).

The following are the features of this encoder and decoder:

1. vendor: Fujitsu, model: IP-700II;
2. video input/output: RCA and S-video;
3. sound input/output: two channel sounds;
4. compression formats:
   (a) 1200 kbps |CBR |SIF |30 Hz |32 kbps(mono);
   (b) 1400 kbps |CBR |HalfD1 |30 H |32 kbps(mono);
   (c) 2000 kbps |VBR |HalfD1 |30 Hz |192 kbps(stereo);
   (d)3000 kbps |VBR |HalfD1 |30 Hz |192 kbps(stereo);
   (e)3000 kbps |VBR |FullD1 |30 Hz |192 kbps(stereo);
   (f)6000 kbps |VBR |FullD1 |30 Hz |256 kbps(stereo);
5. sound frequency: 20 Hz-20000 Hz.

The first item in (4) (a)–(f) shows compression rate for video signal. CBR and VBR stand for constant bit rate and variable bit rate, respectively. SIF stands for source input format. HalfD1 stands for video size of 352 × 480 pixels for one frame. FullD1 stands for 720 × 480 pixels for one frame. 30 Hz means the frame rate of 30 frames per second. The last item means sound compression rate and monophonic sound or stereophonic sounds.

Item (4)-(f) is a standard compression format for standard NTSC broadcasting.

NTSC video signals and stereo sounds from a camcorder are directly inputted into this encoder and a decoder unit can output NTSC video and stereo sound signals directly to a TV monitor.

At the main drill site, Matsumoto Regional Fire Bureau, and Shinshu University Hospital, personal computers are used to decode MPEG-2 compressed video signals created by the encoder aforementioned. These personal computers also have a direct connection to a TV monitor.

We conducted transmission experiments in advance to see video quality differences versus compression rates.

A result showed that there was no big difference in quality between (4)-(c) and (4)-(f). The major difference is the latency of video and sound signal. Item (4)-(c) has about a 0.5 second between an encoder and a decoder. On the contrary, item (4)-(f) has almost no delay. This unit was found to have very high video quality at high compression rates.

Taking this result into account, we decided to use (4)-(c) compression format in this drill.

As a result, transmission rates between Yakushi Park relay station and NPGITC should be more than 4 Mbps when two video streams from the main drill site and the helicopter flow at the same time.

The distance between the main drill site and Yakushi Park relay station requires the same transmission rate.

An IEEE802.11b wireless LAN unit cannot achieve this transmission rate. That is one of the reasons why we developed IEEE802.11g wireless LAN units.

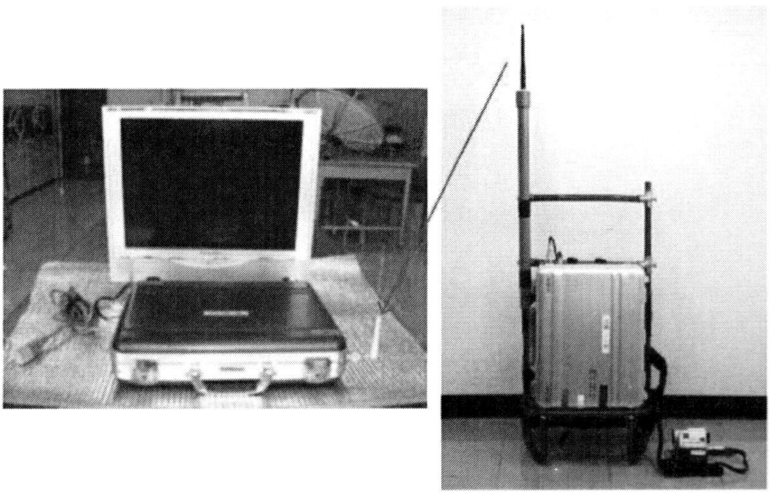

(a) (b)

Figure 11. A PORTABLE LIVE VIDEO TRANSMISSION SYSTEM: (A) VIDEO RECEIVER, (B) VIDEO TRANSMITTER.

On the optical network between NPGITC and Shinshu University Hospital, a compression format of (4)-(f) is used. Therefore, there is almost no delay in video and sound transmission.

3.3. A Portable Live Video Transmission System

Figure 11 shows the portable live video transmission system. Figure 11(a) is a video receiver and Figure 11(b) is a video transmitter.

The transmitter provides direct connection to a camcorder. The receiver also provides direct connection to a TV monitor.

This system uses MPEG-2 as video-compression formats and IEEE802.11b wireless LAN unit for wireless transmission of MPEG-2 compressed video and sound packets. The MPEG-2 compression rate is 2 Mbps as in (4)-(c).

This transmitter can operate for about 3 hours with an internal battery, and weighs about 5 kg.

The coverage of this system is about 200 m on the condition that it is in line of sight using 2.14 dBi dipole antennas at the transmitter and the receiver.

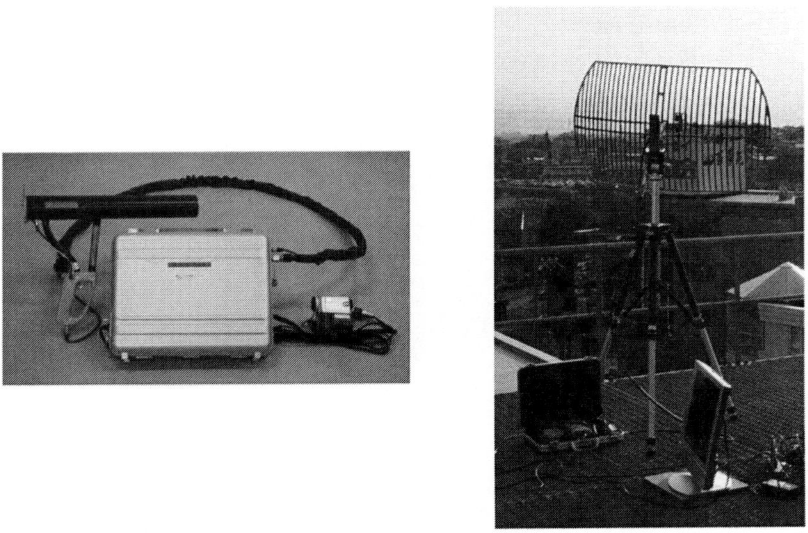

Figure 12. A COMPACT HELICOPTER LIVE VIDEO TRANSMISSION SYSTEM: (A) A VIDEO TRANSMITTER, (B) A VIDEO RECEIVER.

3.4. A Compact Helicopter Live Video Transmission System

Figure 12: shows the compact helicopter live video transmission system. Figure 12(a) is a video transmitter for use in a helicopter. Figure 12(b)shows a video receiver for use on the ground.

It has almost the same configuration as the portable live video transmission system.

The difference is that the video transmitter has a special antenna with a shape of a gun, and that a cable connecting between the antenna and the case, a wireless LAN unit inside the case, and so forth are covered with electromagnetic field shielding fabrics. These fabrics reduce the emission of radio waves and prevent all instruments on the helicopter from malfunctioning by unwanted radio waves.

This system can transfer MPEG-2 compressed video signals and sounds using an IEEE802.11b wireless LAN unit. The video compression rate selected is 2 Mbps.

Figure 13. A MEMBER OF A RESCUE TEAM IN A HELICOPTER.

With this gun-type antenna with a gain of about 17 dBi and a 24 dBi grid parabolic antenna, the coverage is more than 20 km.

This transmitter can operate for about 3 hours with an internal battery, and weighs about 5 kg as well.

Figure 13 shows a member of Nagano Air Rescue team in a helicopter sending live video images using the gun-type antenna.

3.5. Sound Transmission from a Hospital

To enable doctors at Shinshu University Hospital and officials at the Matsumoto Regional Fire Bureau to guide paramedics at the main drill site, we used an IP telephony system or voice over IP system (VoIP).

Figure 14 shows the VoIP unit we used in this drill. This VoIP unit has the following features:

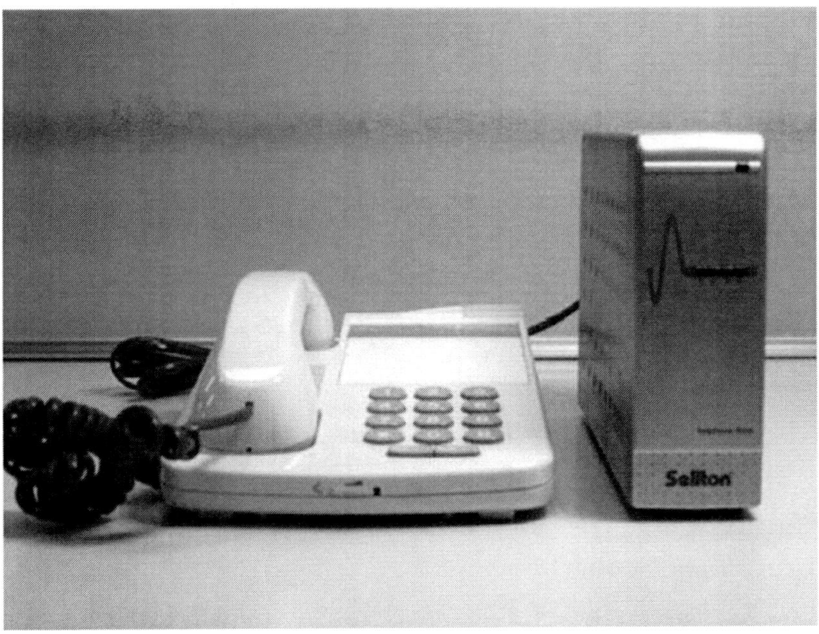

Figure 14. VOIP UNIT AND A TELEPHONE SET.

1. vendor: Soliton Systems Inc.,
2. model: Solphone1204,
3. coding format: G.729ab, PCM (64 kbps),
4. signalling Protocol: H323, Soliton,
5. interface: Ethernet 10Base-T; 1 port, telephone; 4 ports.

This VoIP unit does not require neither session initial protocol (SIP) server nor private branch exchange (PBX) to put into operation. As a result, setting up IP telephony system using this unit is very easy. It can be done by just connecting to the network hub.

We used PCM (64 kbps) codec which has a better sound quality than G.729ab (8 kbps) codec, even though PCM required much more transmission bandwidth.

Figure 15 shows IP telephony system we set up on this wireless LAN network.

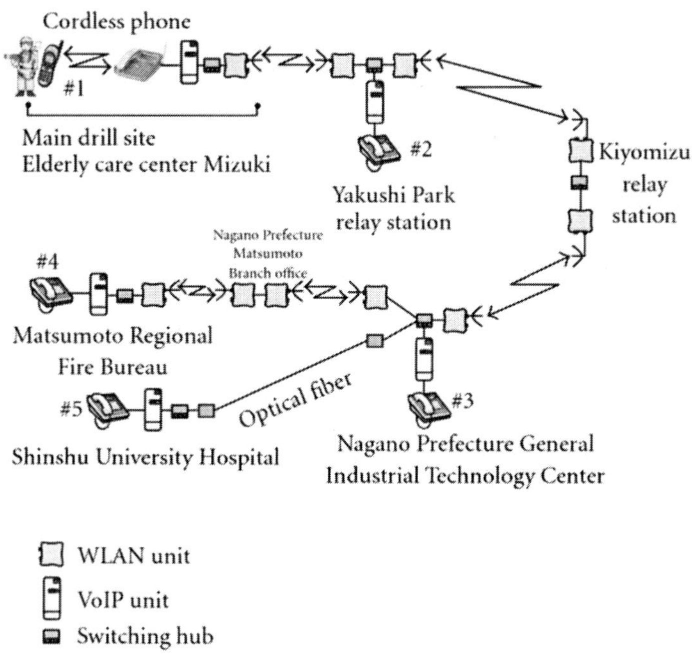

Figure 15. IP TELEPHONY NETWORK.

As is mentioned above, the SIP server and the PBX are not used in this network. A number after # stands for a phone number. We set up each VoIP unit to have a single-digit speed-dial number.

With this system, doctors and officials can call paramedics and talk with them by just speed dialling a phone number. Cordless phones were used at the main drill sites to enable firefighters to communicate with doctors at the hospital, while still being able to remain mobile.

3.6. Equipment Used in Each Place

Images from Figure 16 to Figure 21 show equipment used for the main drill site, each relay station, NPGITC, the Matsumoto Regional Fire Bureau, and Shinshu University Hospital.

Figure 16. EQUIPMENT AT THE MAIN DRILL SITE: (A) WIRELESS LAN UNIT AND ITS ANTENNA; (B) VIDEO MONITORS FOR A DECODER FOR THE PORTABLE LIVE VIDEO TRANSMITTER, A COMPACT HELICOPTER LIVE VIDEO TRANSMISSION SYSTEM, AND AN ENCODER FOR SHINSHU UNIVERSITY HOSPITAL AND MATSUMOTO REGIONAL FIRE BUREAU.

At the main drill site and the Mizuki relay station, flat panel antennas were used, since the distance between these two places is only about 4 km (Figures 16 and 17).

Between the Mizuki relay station and the Kiyomizu relay station, 24 dBi grid parabolic antennas were used (see Figures 17 and 18). Grid parabolic antennas with a gain of 24 dBi were also used between the Kiyomizu relay station and NPGITC (see Figures 18 and 19).

Figure 17. YAKUSHI PARK RELAY STATION.

Figure 18. KIYOMIZU RELAY STATION.

Figure 19. NAGANO PREFECTURE GENERAL INDUSTRIAL TECHNOLOGY CENTER (NPGITC).

At the Matsumoto Regional Fire Bureau, MPEG-2 compressed video image data were decoded by personal computer and displayed with a flat panel display (Figure 20). A VoIP unit is not shown in Figure 20, but it was used.

Figure 20. MATSUMOTO REGIONAL FIRE BUREAU.

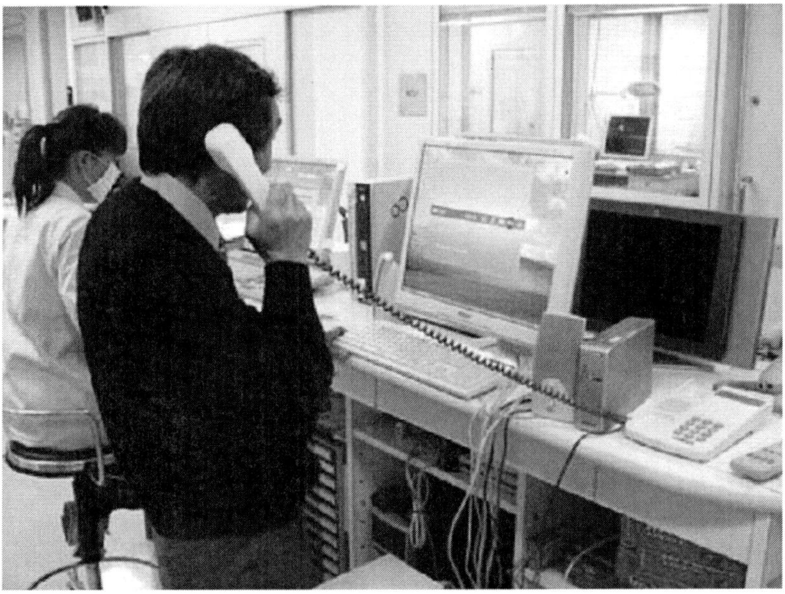

Figure 21. SHINSHU UNIVERSITY HOSPITAL.

At Shinshu University Hospital, the MPEG-2 decoder used with a personal computer and VoIP unit to give medical instructions to paramedics were used (Figure 21).

4. EXPERIMENT RESULTS

4.1. Developed IEEE802.11g Complied Wireless LAN Unit

Prior to this drill, we conducted two measurements on the maximum transmission rates.

The first was to measure maximum transmission rate in an ideal environment, in which there were no interferences. This measurement configuration is shown in Figure 22. Two wireless LAN units are connected to each other via a fixed attenuator and a variable attenuator.

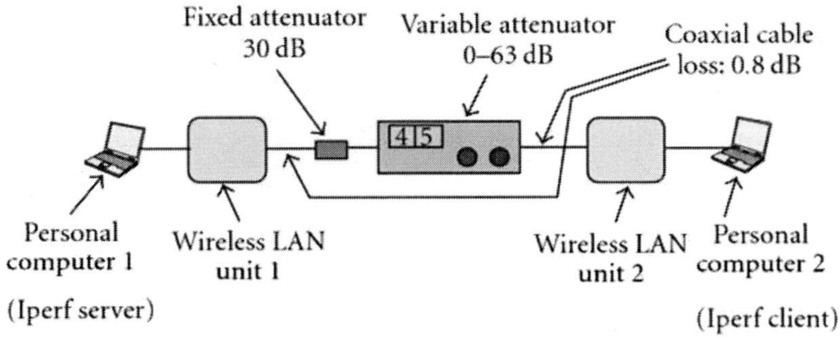

Figure 22. MEASUREMENT SETUP.

To measure the maximum transmission rate, software called "Iperf" was used. This software can measure the maximum transmission of a network. This software was installed onto 2 different personal computers. One of these computers was to be set up to run as an Iperf server and the other one was set up to run as Iperf client. The maximum transmission rate was measured for 30 seconds.

Measurement results showed that the maximum transmission rates on average were 20.2 Mbps.

An actual maximum transmission rate measurement in the open air was also conducted in advance.

In this measurement, two wireless LAN units were placed at a distance of about 45 km in line of sight. Each wireless LAN unit used a grid parabolic antenna with a gain of 24 dBi as shown in Figure 4.

The measured maximum transmission rate was 6.6 Mbps and the maximum transmission rate on average in 30 seconds was 4.6 Mbps.

4.2. High-Quality Live Video Image Transmission

Figure 23 shows a firefighter at the main drill site who is sending video images by the portable video transmission system. A video image example taken and sent by this system is shown in Figure 24.

Figure 23. THE PORTABLE LIVE VIDEO TRANSMISSION SYSTEM.

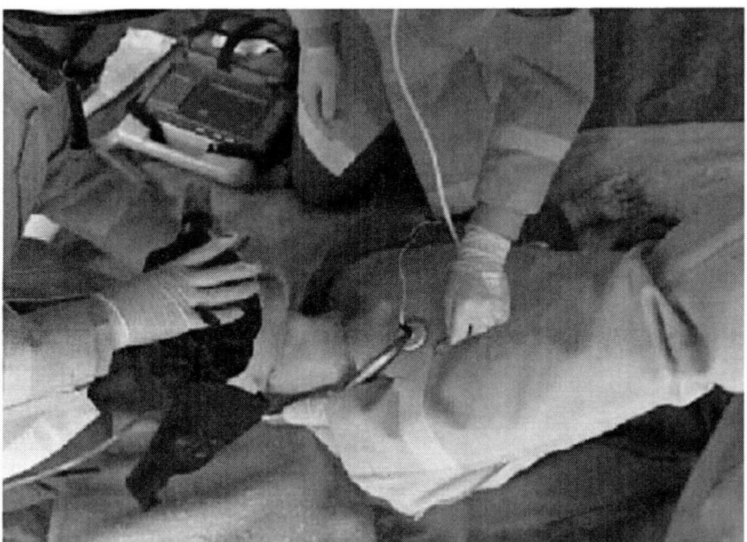

Figure 24. TRANSMITTED IMAGE EXAMPLE OF EMERGENCY TREATMENT DRILL WITH DUMMY.

The distance between the base station and the portable video transmitter was about 50 m. The received picture had quality almost as good as TV broadcasting images.

Figure 25 shows an example of the image received at the main drill site transmitted by the portable video transmission system.

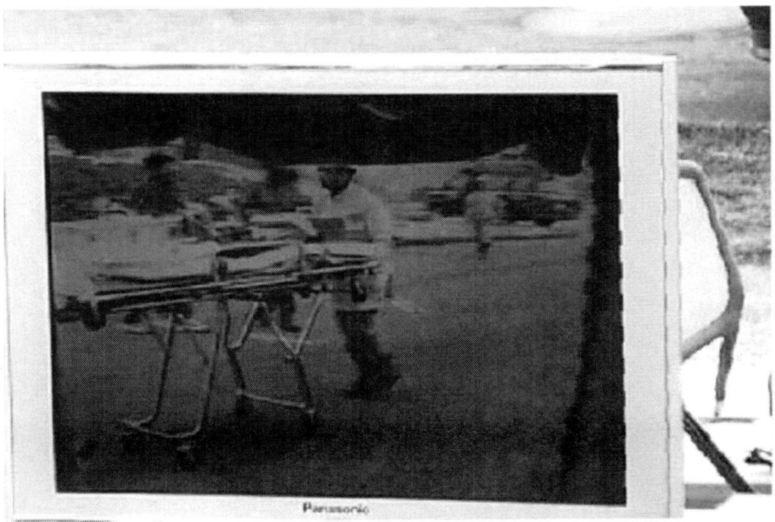

Figure 25. RECEIVED VIDEO IMAGE EXAMPLE FROM THE PORTABLE LIVE VIDEO TRANSMISSION SYSTEM AT THE MAIN DRILL SITE.

These images had high quality and were received without problems, such as stoppage for short periods in video transmission.

Figure 26 shows an example of the image received at the main drill site transmitted by the compact helicopter live video transmission system.

These images had higher quality than a professional system, which uses spectrum of 15 GHz licensed only for disaster-related organizations in Japan. This system transmits video signals using analog modulation technology.

Figures 27 and 28 show video images received over our network at Shinshu University Hospital. An emergency-treatment doctor is watching and giving instructions to paramedics using IP telephony system in Figure 27. This picture shows the received image quality is almost equal to TV broadcasting pictures.

Figure 26. RECEIVED VIDEO IMAGE EXAMPLE FROM THE COMPACT HELICOPTER LIVE VIDEO TRANSMISSION SYSTEM AT THE MAIN DRILL.

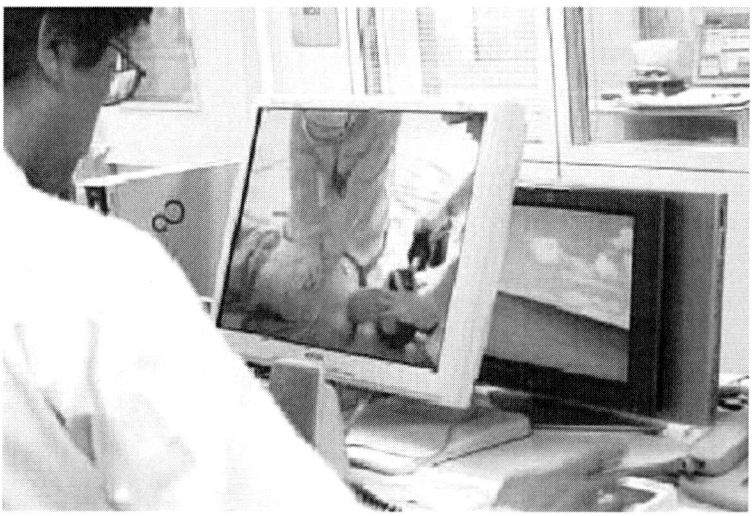

Figure 27. A RECEIVED VIDEO IMAGE EXAMPLE AT SHINSHU UNIVERSITY HOSPITAL SENT FROM THE MAIN DRILL SITE.

Figure 28. A RECEIVED VIDEO IMAGE EXAMPLE AT SHINSHU UNIVERSITY HOSPITAL SENT FROM THE RESCUE HELICOPTER.

This system has about a one-second delay in the video image viewing between the drill site and Shinshu University Hospital, because of the compression and decompression delay of the video encoder and decoder.

Figure 28 shows a video image taken and sent from the rescue helicopter. This system can give clear live views of the devastated area.

The link speeds of wireless LAN units between each site were as follows:

1. the main drill site and Mizuki relay station: 54 Mbps;
2. Mizuki relay station and Kiyomizu relay station: 36 Mbps;
3. Kiyomizu relay station and NPGITC: 54 Mbps.

An IP telephony system was able to be operated with this video transmission system at the same time and gave clear sound quality to both the paramedics and the doctor.

The setup of all makeshift relay stations can be completed and operational in about 2 hours, thus giving our network an extremely fast deploy time anywhere in the world.

5. DISCUSSION

Live video images obtained at Shinshu University Hospital have extremely exceptional quality as found in today's broadcasting fields. This enables doctors to check patients' complexion and health condition.

This live video image transmission is likely to cut about 1 hour off the time at which a surgery starts after a patient is transported from this drill site to Shinshu University Hospital, whereas without it, it could take up to 3 hours to start surgery.

About a one-second delay of video and sound occurs, due to encoding and decoding of video images 3 times.

The reason why the link speed between Mizuki relay station and Kiyomizu relay station was 36 Mbps is that antennas with a gain of 24 dBi were used. This wireless LAN unit has been designed to be used with an antenna with a gain of 26.5 dBi in order to achieve a wireless connection of 30 km at a transmission rate of 54 Mbps.

6. CONCLUSION

We have developed IEEE802.11g complied wireless LAN units and conducted experiments to confirm the capability of these units in an earthquake drill organized by a regional fire bureau.

Features of this drill were to send high-quality video images and sound of areas struck by a large-scale and strong earthquake to hospitals, and to give instructions to paramedics by doctors, using a makeshift wireless LAN network made up of these LAN units.

In this drill, a portable live video transmission system and a compact helicopter live video transmission system we developed were also used successfully in the main drill site area. These systems consisted of IEEE802.11b wireless LAN units we had already developed.

The networks that we established and successfully operated consisted of only IEEE802.11b/g wireless LAN units.

Conclusions are as follows.

1. The experiment conducted in the drill shows that developed wireless LAN units have a capability of transmitting data 41 km with a link speed of 36 Mbps using 24 dBi parabolic antennas. This distance established a record as the longest distance MPEG-2 compressed video and audio information has been transmitted over a wireless LAN network in Japan.
2. We succeeded in transmitting MPEG-2 compressed video images at 2 Mbps over the temporally wireless LAN network we established for this drill. Images were successfully sent from a portable live video transmission system and a compact helicopter transmission system to Shinshu University Hospital.
3. MPEG-2 compressed images, sent at 2 Mbps using the encoder and decoder we used, have a high quality enough for use in diagnosing patient conditions.
4. VoIP telephony system we used is easy to set up and can give clear sound quality. This system is very helpful in establishing communication where any type of telephone system is knocked out of service.
5. The earthquake drill conducted shows that live video images transferred over this wireless LAN network have enough quality for use in telemedicine in emergency treatment.

Use of a helicopter video feed, we believe, would enable emergency personnel including doctors to better assess a situation. This could help with decisions relating to the amounts of aid needed, time factors in dangerous circumstances, and where best to locate rescue crews.

We believe that our portable units would be an asset in any emergency situation, anywhere in the world.

Our wireless LAN units are easy to set up, operate, and maintain, and are able to transmit real-time images of a disaster area. These units would be a great benefit for emergency units and disaster-relief organizations in that they would open up great opportunities to put aid where it is needed most.

ACKNOWLEDGMENTS

This paper is a part of the project of the development of long-range, high-speed wireless LAN to establish an emergency network for the disaster prevention and the postdisaster communications supported by the Fire and Disaster Management Agency of Japan (FDMA). The authors wish to give their thanks to FDMA.

Authors' Affiliations

(1) Information Technology Department, Nagano Prefecture General Industry Technology Center

(2) Telecommunication Network Systems Engineering Department, Mistubishi Cable Insudtries Ltd.

(3) Rescue and Ambulance Department, Matsumoto Regional Fire Bureau

(4) Nagano Air Rescue Team, Nagano Prefecture Emergency Management, Nagano Prefecture

(5) Department of Medical Informatics, Shinshu University Hospital

REFERENCES

1. Fujimori K, Hareyama M: Teleradiology with a broadband network. Proceedings of the 22nd Joint Conference on Medical Informatics (JCMI '02), Supplement, November 2002, Kawasaki, Japan 81-82.
2. Garshnek V, Burkle FM: Applications of telemedicine and telecommunications to disaster medicine: historical and future perspectives. Journal of the American Medical Informatics Association 1999, 6(1):26-37. 10.1136/jamia.1999.0060026
3. Miyashita T, Takizawa M, Nakai K, et al.: Telemedicine of the heart—real-time telescreening of echocardiography using satellite telecommunication. Circulation Journal 2003, 67(6):562-564. 10.1253/circj.67.562
4. Cermack M: Monitoring and telemedicine support in remote environments and in human space flight. British Journal of Anaesthesia 2006, 97(1):107-114. 10.1093/bja/ael132

5. Yoshida A, Date T, Sasaki H, et al.: Telemedicine for rural area and islands using satellite internet access. Japanese Journal of Telemedicine and Telecare 2006, 2(2):162-163.
6. Kanthraj GR, Srinivas CR: Store and forward teledermatology. Indian Journal of Dermatology, Venereology and Leprology 2007, 73(1):5-12. 10.4103/0378-6323.30643
7. Koanantakool T: Emergency and educational communication vehicle (EECV). ITU/ESCAP Regional Workshop on Disaster Communications, December 2006, Bangkok, Thailand
8. Simmons SC, Murphy TA, Blanarovich A, Workman FT, Rosenthal DA, Carbone M: Telehealth technologies and applications for terrorism response: a report of the 2002 coastal North Carolina domestic preparedness training exercise. Journal of the American Medical Informatics Association 2003, 10(2):166-176. 10.1197/jamia.M1205
9. Sadiq MA, Usman K, Juzoji H, Igarashi K, Tanaka K, Nakajima I: Wireless IP network for emergency disaster telemedicine: case study of Yogyakarta-central java earthquake. Japanese Journal of Telemedicine and Telecare 2006, 2(2):128-129.
10. Pattichis CS, Kyriacou E, Pattichis MS, Panayides A, Pitsillides A: A review of m-health e-emergency systems. Proceedings of the International Special Topic Conference on Information Technology in Biomedicine (ITAB '06), October 2006, Ioannina, Greece
11. Donahoo M, Steckler B: Emergency wireless networks. Proceedings of IEEE Military Communications Conference (MILCOM '05), October 2005, Atlatnic City, NJ, USA 4: 2413-2420.
12. Cordova H, Boets P, Van Biesen L: Analysis and applications of Wi-MAX standard: forecasting the trends.
13. Crouse J, D'alessandro M, Lengerich F: WiMAX and the future of wireless technology. 2007.
14. Neves P, Simoes P, Gomes A, et al.: WiMAX fir emergency services: an empirical evaluation. Proceedings of the International Conference on Next Generation Mobile Applications, Services and Technologies (NGMAST '07), September 2007, Cardiff, UK 340-345.
15. Chan TC, Killeen J, Griswold W, Lenert L: Information technology and emergency medical care during disasters. Academic Emergency Medicine 2004, 11(11):1229-1236. 10.1111/j.1553-2712.2004.tb00709.xView

16. Light J, Bhuvaneshwari A: Mobile infrastructure for emergency medical services. Proceedings of the IASTED International Conference on Telehealth (TELEHEALTH '05), July 2005, Banff, AB, Canada 23-28.
17. Dilmaghani RB, Manoj BS, Rao RR: Emergency communication challenges and privacy. Proceedings of the 3rd International Conference on Information Systems for Crisis Response and Management (ISCRAM '06), May 2006, Newark, NJ, USA 1-9.
18. Arisoylu M, Mishra R, Rao R, Lenert LA: 802.11 wireless infrastructure to enhance medical response to disasters. Proceedings of the American Medical Informatics Association Symposium (AMIA '05), October 2005, Washington, DC, USA 1-5.
19. McCurdy NJ, Griswold WG, Lenert LA: Realityflythrough: enhancing situational awareness for medical response to disasters using ubiquitous video. Proceedings of the American Medical Informatics Association Symposium (AMIA '05), October 2005 510-514.
20. Nakamura M, Takizawa M, Murase S, et al.: Study on emergency medical support for mountain huts in Japan Alps over wireless LAN and CATV network. Proceedings of the 21st Joint Conference on Medical Informatics (JCMI '01), November 2001, Kawasaki, Japan 621-622.
21. Nakamura M, Shoshin K, Yuying Y, Yutaka M, Masaomi T: Telemedicine for mountain climbers with high quality video image transmission on Japan Alps. Proceedings of the 22nd Joint Conference on Medical Informatics (JCMI '02), November 2002, Kawasaki, Japan 176-177.
22. Nakamura M, Yang Y, Kubota S, et al.: Experimental telemedicine network with the long range wireless LAN and CATV for mountain climbers in the Japanese Alps. Proceedings of the IASTED International Conference on Telehealth (TELEHEALTH '05), July 2005, Banff, AB, Canada 29-36.

3
CHAPTER

LINK LAYER CORRECTION TECHNIQUES AND IMPACT ON TCP'S PERFORMANCE IN IEEE 802.11 WIRELESS NETWORKS

Purvang Dalal[1], Mohanchur Sarkar[2], Kankar Dasgupta[3], Nikhil Kothari[1]

ABSTRACT

TCP performance degrades when end-to-end connections extend over wireless links which are characterized by high Bit Error Rate and intermittent connectivity. Such degradation is mainly accounted for TCP's unnecessary congestion control actions while attempting TCP loss recovery. Several independent link loss recovery approaches are proposed by researchers to reduce number of losses visible at TCP. In this paper we first presented a survey of loss mitigation techniques at wireless link layer. Secondly performance evaluation for TCP through Type 0 Automatic Retransmission Request mechanism in erroneous Wireless LAN is presented. In particular, simulations are performed taking into account the wireless errors introduced over IEEE 802.11 link using a well-established 2-State Markov model. TCP performance is evaluated under different settings for maximum link retransmissions allowed for each frame.

Simulation results show that, link retransmission improves TCP performance by reducing losses perceived at TCP sender. However, such improvement is often associated with adverse effect on other TCP parameters that may cost a lot in return under extreme network conditions. In this paper an attempt is made to observe impact of link retransmissions on the performance of multiple TCP flows competing with each other. The analysis presented in this paper signifies the scope for maximizing TCP's throughput at the least possible cost.

Keywords: Wireless LAN, Cross-Layer Scheme, Medium Access Control, Round Trip Time, Bandwidth Estimation

1. INTRODUCTION

In recent scenario, the IEEE 802.11 [1] or 3G/4G wireless networks represent a significant milestone in the provisioning of "anywhere - anytime" Internet connectivity to the wandering users [2] [3] . Together, Transmission Control Protocol (TCP) [4] has remained as the dominant transport layer protocol for majority of Internet applications [2] [5] . Unlike wired links, wireless links have high, bursty and random errors due to atmospheric conditions, terrestrial obstructions, fast and multi-path fading, active interference, mobility and resource constraint [3] [6] . Consequently, significant amount of efforts [7] [8] have been devoted to the provisioning of reliable TCP delivery for a wide variety of applications over different wireless infrastructures.

Impact of the losses on TCP performance has been characterized in [9] using the following Equation (1). From the view of TCP, the throughput is inversely proportional to the packet losses perceived by TCP sender.

$$B(p) \approx \left(\frac{cwnd}{RTT}, \frac{1}{RTT\sqrt{\frac{2bp}{3}} + T_0 \min\left(1, 3\sqrt{\frac{3bp}{8}}\, p\left(1+32p^2\right)\right)} \right) \tag{1}$$

Nevertheless, when the losses are seen at TCP sender due to reasons other than network congestion, end-toend throughput is compromised due to its loss recovery attempts with detrimental transmission rate. Therefore it is advocated to endeavor initial loss recovery at link layer (link recovery) and to attempt end-to-end TCP recovery [10] [11] , only if inescapable. Attempts to make wireless links resemble wired ones for high level protocols are reflected in several approaches operating at a link layer.

In this paper, some of the prominent approaches made at wireless link layer for improving TCP's performance are discussed first. In the modern Wireless LAN (WLAN), due to the technology progress, the physical channel condition is comparatively good enough to provide low Packet Error Rate (PER) and consequently higher datarate [2] [12] . On the other hand, to guarantee the correctness of control information, the control data rate increases very slowly (e.g. in IEEE 802.11 g data-rate is raised up to 54 Mbps and control data-rate is increased up to 6 Mbps only). In this situation, the overhead of control information including the link layer acknowledgement (ACK) increases significantly. The above grounds for very low efficiency of Automatic Retransmission reQuest (ARQ) protocol. As a result, the effect of link recovery in such networks should be thoroughly analyzed. In earlier work [13] , the performance of a series of ARQ mechanisms are analyzed in the link layer without considering the TCP layer effect. In this work, a set of simulations are performed and results are analyzed to show impact of link recovery on TCP's performance. Analysis presented in this paper not only addresses the merits and demerits of link recovery attempts but also give a new direction towards enhancing TCP's performance further.

Our results show that with light errors on the IEEE 802.11 link, better TCP performance is achieved using link layer attempts. Since, link recovery attempt increases the Round Trip Time (RTT) estimate at TCP, it may adversely effects on the TCP performance by deteriorating the transmission rate at TCP sender (It is controlled using TCP's internal parameters congestion window (cwnd) and RTT). In fact, even with the use of link recovery, sub-optimum TCP's performance is seen exclusively in the extreme network environments; wherein total loss recovery is not attained at link layer. The rest of the sections are as follows. Section 2 discusses the pure link layer proposals designed to mitigate wireless errors. Section 3 provides a brief summary of link recovery mechanism over IEEE 802.11 links and Section 4 discusses the simulation results to show its influence on the end-to-end TCP's performance in wireless scenario. Section 5 concludes this paper.

2. PURE LINK LAYER ENHANCEMENTS

Addressing link errors near the site of their occurrence appears intuitively attractive for several reasons.

1) Link layer approaches are likely to respond more quickly to changes in the error environment and therefore local recovery over a point to point link is more efficient than end-to-end TCP recovery [10] .

2) Since, local recovery mechanism commonly operates on exactly the link that require its rendering, the deployment of new and existing wireless link protocol is more feasible than applying novel transport layer solution.

3) The link recovery approaches do not violate the modularity of the protocol stack [13] and in this context they are different from Cross-layer [14] approaches.

Various approaches have been proposed at link layer to optimize the performance of TCP in wireless networks. These approaches are broadly classified into two groups on the basis of the awareness of the transport layer protocol; (a) TCP unaware approaches and (b) TCP aware approaches.

2.1. TCP Unaware Approaches

IEEE 802.11 and Cellular networks employ TCP unaware link and physical layer mechanisms with an objective to reduce the PER to a level that would not cause significant performance degradation at TCP. In the above networks, physical layer achieves high coding gain with the use of convolution and turbo coding [15] . Additionally interleaving provides time diversity as a further safeguard against burst errors. The most remarkable TCP unaware link layer implementations in the above wireless technologies uses ARQ mechanism and packet scheduling to reduce effect of losses on TCP's performance. For example, Selective Repeat ARQ with scheduling protocols such as RLP and RLC are used in 3G1X and UMTS, respectively [15] . On the other side, IEEE 802.11 uses stop and wait ARQ [13] alone to recover from transmission losses locally. ARQ is a closed-loop mechanism that requires feedback and retransmission and is invoked when packets containing bit errors are discarded. Link ARQ consumes additional network resources only when a packet is retransmitted. The mechanism generally operates more efficiently for low bit rates. On a downside, the above mechanisms may cause delay variability and out-of-order delivery of packets at TCP. This can result in duplicate retransmissions from the two layers and hence inefficient utilization of the network capacity. An undesirable side effect of ARQ is that it may interfere with independent TCP mechanisms. Forward Error Correction (FEC) is a popular error mitigation mechanism for detection and recovery of transmission losses [16] without retransmissions, which is critical for lossy links exhibiting long delays. Unlike ARQ, FEC doesn't interfere with TCP mechanisms. FEC suffers from dead-weight overheads in favorable conditions, resulting in a waste of limited bandwidth. Furthermore FEC requires additional resources (processor, memory) and power consumption.

Based on the recommendations made by IETF and 3GPP, most of the link layer techniques for wireless lossy links, involve making the link layer reliable using hybrid FEC/ARQ [17] . Formerly a good summary of hybrid FEC/ARQ techniques for the link layer is provided in [13] . In a Type 0 scheme, recovery for losses is offered only through ARQ. This technique is applicable to links having losses subject to infrequent interference. In Type I technique, initial transmission is protected using FEC. If this fails, ARQ is used to repeat the transmission. The main characteristic of Type II schemes is that ARQ is used first followed by FEC. Typically the initial transmission has error detection built in but no error correction. Though hybrid ARQ/FEC is better than either FEC or ARQ alone, its performance also significantly degrades for higher loss rates despite unreasonably high amounts of ARQ retries, fragmentation of IP packets, FEC overhead and buffering [15] . Asymmetric Reliable Mobile Access In Link Layer (AIRMAIL) and Transport Unaware Link Improvement Protocol (TULIP) are the known link layer implementation based on TCP unaware link layer techniques as discussed in [7] .

Barakat et al. [11] reported that ARQ mechanisms may vary the characteristics of the network, affecting the functionality of the upper layer protocols. Moreover lack of knowledge of the protocol operating at the transport level may results into end-to-end performance degradation in presence of independent link layer mechanisms. For instance, an approach without awareness of the transport protocol may cause local link layer retransmission of a packet, as well as duplicate acknowledgement (dupack), since retransmissions can be performed on both layers. This led to significant efforts in development of TCP aware link layer approaches [18] .

2.2. TCP Aware Approaches

The link layer approaches in which enhancements tailored for wireless environments are made known to TCP are broadly referred as TCP aware link approaches. A good overview of these approaches can be found in [7] [17] [18] . The representative mechanisms employed in various approaches include Snooping, Delayed Acknowledgement and Performance Enhancement Proxy (PEP).

2.2.1. Snooping Mechanism

Most of the approaches which operate at this level rely on some intermediate point within the end-to-end connection for the introduction of performance improvement. In Snoop protocol [19] , agent located at the Base Station (BS) monitors every packet that passes through the TCP connection in both directions. The Snoop agent suppresses the dupacks for lost TCP packets and retransmits locally, thereby preventing unnecessary invocation of congestion control mechanism by sender. Other enhancements proposed in similar category are TCP SACKaware Snoop Protocol and SNACK-NS (New Snoop) [7] . This protocol may cause additional delay for TCP packets, which results into futile TCP retransmissions at TCP sender over the slower wireless link. However, the problem can be solved by minimizing the number of retransmissions at the link level [7] .

WTCP [20] , although operating in a similar way, introduces more accurate RTT estimation and thus preventing reduction in TCP throughput. WTCP conceals the difference in RTT from the sender, thus avoiding needless timeout caused by reason of local retransmissions. Both Snoop and WTCP must, however, have access to the header of TCP packets in order to function, which reduces their usefulness value if traffic is encrypted. The use of Snoop agent is also exploited in wireless TCP refinements with split connection schemes [7] . Snoop and WTCP increase complexity at BS, especially when transport layer per flow support is required. TCP PEPs [21] are TCP aware link layer mechanisms for handling bandwidth asymmetry, TCP aware FEC provisioning or adaptive link frame sizing [22] . Split connection approaches [7] and PEPs, though effective in many cases, break TCP end-to-end semantics and its modularity and hence not considered as pure link layer approaches.

2.2.2. Delayed Acknowledgements

Another class of work includes Delayed Duplicate Acknowledgment (DDA) [23] , which preferred particularly when IP encryption is used. With a view to reduce interference between the TCP and link retransmissions, TCP receiver delays the third and subsequent dupacks for some interval "d". This provides enough time for link layer to recover the lost packet using local retransmission, and to prevent TCP sender from fast retransmit. Different from the Snoop protocol, in DDA dupacks are not dropped immediately but rather delayed a certain length of time. Delayed Acknowledgements over Wireless Link (DAWL) [24] tries to simplify the system by introducing modifications only in the ARQ scheme at the link layer and does not consider local retransmission at the BS. This design is advantageous in case of a BS crash [7] . However, in presence of congestion, the performance of these schemes is degraded, as the essential fast retransmissions are unnecessarily delayed.

Since different link layer technologies have diverse capabilities for error recovery, there is no de-facto standard for link layer protocols. All link layer approaches try to reduce the effect of erroneous wireless links on the performance of TCP, but do not completely shield the TCP sender from all types of wireless errors particularly in the event of lengthy disconnection or high Bit Error Rate (BER). The main advantage of link layer approaches is the maintenance of end-to-end semantics, without modification of higher protocol layers. This makes it possible to leave untouched the existing implementations of the protocol stack in the various operating systems.

3. IEEE 802.11 LINK RECOVERY MECHANISM

The basic channel access method in 802.11 networks is the Distributed Coordination Function (DCF) in which 802.11 Medium Access Control (MAC) layer uses the Carrier Sense Multiple Access/Collision Avoidance (CSMA/CA) mechanism before transmitting any frame. The IEEE 802.11 MAC layer employs Type 0 ARQ concept for loss recovery with the contention based transmission policy. The ARQ is the only error control method specified in the standard and no FEC coding is used as stated earlier.

Over an IEEE 802.11 link, whenever a wireless node notices unsuccessful transmission for a frame, it attempts local retransmission of a frame in accordance with RetryLimit (RL) [1] . To explain, as shown in Figure 1, a wireless node after sending a frame waits for a positive ACK from wireless receiver. The sender attempts retransmission of the frame whenever ACK is not received before expiration of ACK timeout. The node is allowed to attempt retransmissions for a particular frame maximum up to RL. Then after node initiates for new transmission and the previous frame is considered to be lost. Thereby the upper layer is made responsible for recovery of the frames lost after maximum retransmissions.

Such link retransmissions additionally delay the transmission of subsequent TCP packet residing at Interface Queue (IFQ). This additional delay (T_{ARQ}) is approximated using Equation (2).

$$T_{ARQ} = n \cdot \left(t' + T_{CONT}\right) \tag{2}$$

T_{ARQ} increases with (a) increase in the number of unsuccessful retransmissions (n) and (b) increase in link delay (t'). Here T_{CONT} signifies the contention delay over a shared medium [1] . TCP's RTT estimation includes wired link delay and processing delay at nodes (the combined is referred as RTT_0), delay due to network con gestion (d_c) and T_{ARQ} as shown in Equation (3).

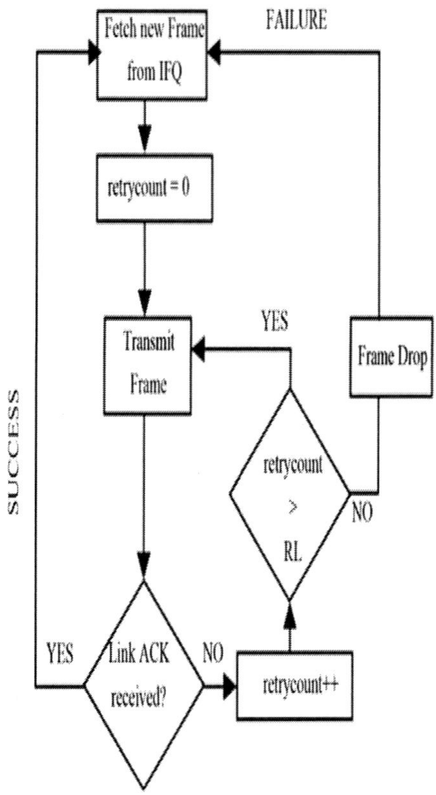

Figure 1. IEEE 802.11 link loss recovery mechanism.

$$\mathrm{RTT} = \mathrm{RTT}_0 + d_c + T_{ARQ} \tag{3}$$

In short, link recovery mechanisms shield TCP from the wireless losses and hence prevent undue reduction of cwnd, but at the same time increase RTT estimation at TCP sender. The earlier effect prevents sacrificed transmission rate and the later accounts for diminished growth in transmission rate. In the next section, an analysis of results obtained through simulations is presented to illustrate impact of link recovery on TCP's performance.

4. SIMULATION RESULTS AND ANALYSIS

4.1. Simulation Scenario

A network topology illustrated in Figure 2 was used for analyzing Medium Access Control (MAC) and TCP behavior using a TCP connection between sender (S1) and receiver (D). Wireless losses over an IEEE 802.11b link between S1 and BS were introduced using 2-State Markov model. The queue length at intermediate nodes is set in accordance to Bandwidth Delay Product (BDP) in the network.

A TCP/FTP flow was introduced for 100 sec. To investigate the impact of link recovery on the efficiency of link and transport layer protocols, the simulations were performed using different value for RL and Frame Error Rate (FER) in the range mentioned along with results. TCP's RTT estimation and layer efficiency are the measurable parameters used for performance evaluation. Layer efficiency is defined as the ratio of total number of successful transmissions to the total transmissions made by that layer during simulation interval. Nevertheless the most critical parameter under observation is (cwnd/RTT). For better understanding, analysis of results is presented with different traffic conditions; (a) a single TCP flow and (b) multiple TCP flows competing over a bottleneck link. In the mentioned topology, TCP New Jersey [25] (TCP NJ) has demonstrated better throughput compared to that achieved using other well-known TCP variants (i.e. NewReno, Vegas and Westwood [26]). Therefore the analysis is presented for TCPNJ as TCP variant. However similar trend in the results is observed with other TCP variants as well.

4.2. Analysis with Single TCP Flow

Figures 3-8 present analysis based on the results obtained from the experiments.

As shown in Figure 3(a), with link recovery (RL = 7), increase in FER resulted into increase in link retransmissions in order to combat transmission losses locally. Pl. note that even with 0% FER (introduced using error model), losses were witnessed due to channel contention (due to transmission of TCP data and acknowledgement packets on the same channel in reverse direction). It led to noticeable MAC retransmissions for recovering from such losses. During simulations with link recovery (RL = 7), % of wireless TCP drops were significantly

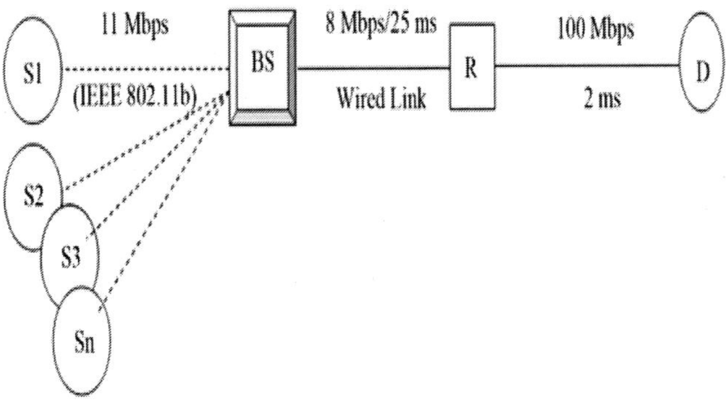

Figure 2. Network Topology.

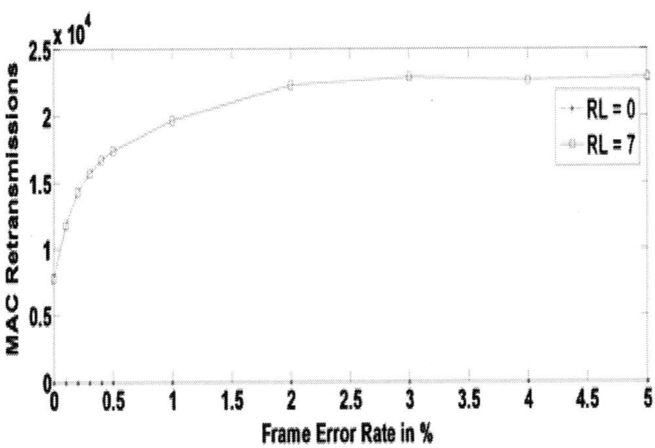

(a)

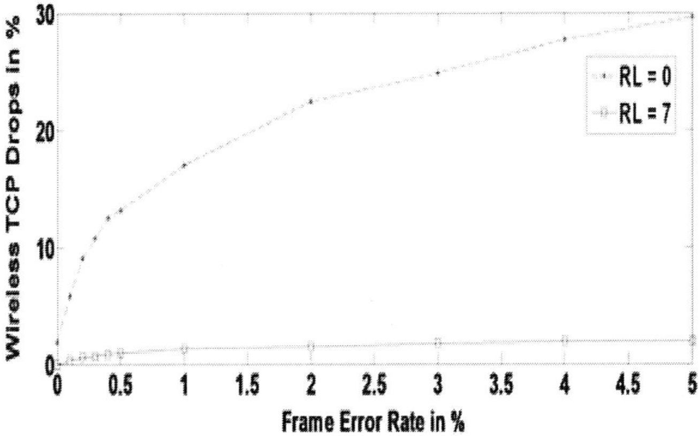

(b)

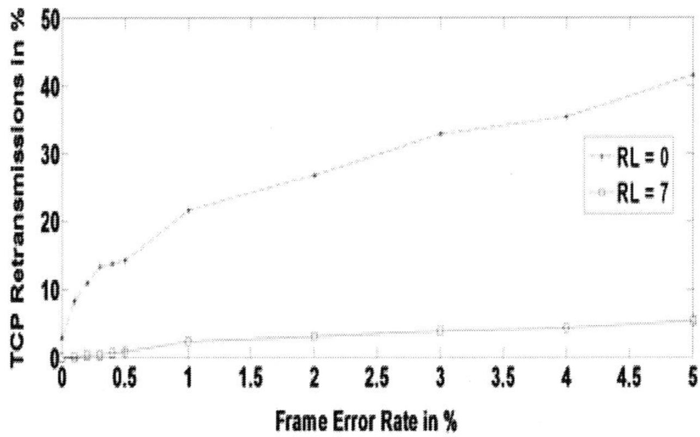

(c)

Figure 3. (a) Impact of link ARQ on MAC retransmissions, (b) Impact of link ARQ on Wireless TCP Drops in %, (c) Impact of link ARQ on TCP Retransmissions (in %).

reduced together with increase in FER compared to those witnessed during simulations without link recovery (RL = 0) (refer Figure 3(b)). Since, the error model introduces losses in % of total transmissions; in order to demystify the impact of link loss recovery correctly the analysis is presented in terms of % of TCP drops. Reduction in wireless TCP drops has caused significant reduction in end-to-end TCP retransmissions (analysis is presented in % of total TCP transmission due to similar reasons stated earlier), as shown in Figure 3(c).

The simulations were performed using single TCP flow and therefore the packet losses were attributed only to the transmission errors on wireless link. During simulations TCP flow encountered timeouts due to either loss of retransmissions or due to loss of TCP acknowledgement in presence of errors on the wireless links, those are very costlier to TCP. Figure 4 represents comparison for number of TCP timeouts with RL = 0 and RL = 7. As seen from the Figure, number of timeouts were considerably condensed when link retransmissions were enabled (i.e. RL = 7).

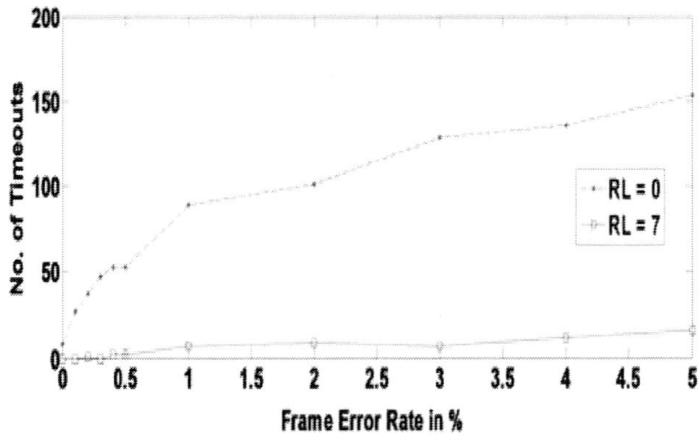

Figure 4. Impact of link ARQ on TCP timeouts.

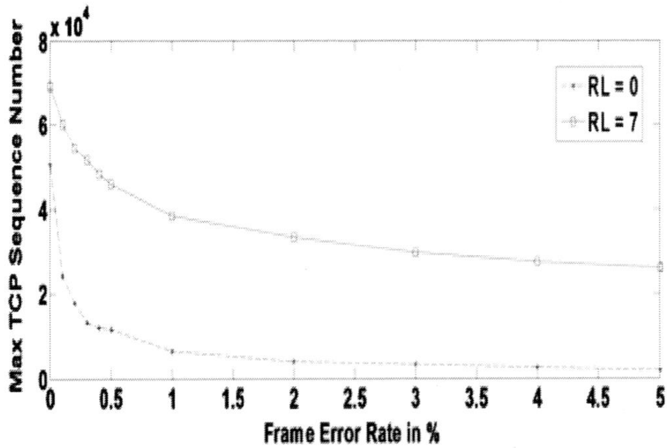

(a)

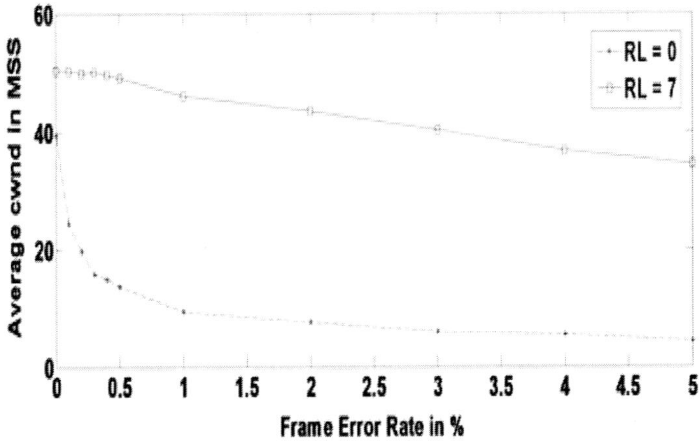

(b)

Figure 5. (a) Impact of link ARQ on TCP's maximum sequence number, (b) Impact of link ARQ on congestion window.

In absence of any support for loss discrimination, TCP reacts to timeout with drastic reduction in cwnd, which is inappropriate for wireless losses. Figure 5(a) shows that without link recovery (RL = 0), TCP's maximum sequence number dropped rapidly with increase in FER; since loss of TCP segments has reduced the cwnd and slowed down TCP's transmissions per RTT. However with link recovery, significantly higher value of TCP's maximum sequence number is obtained, relative to that obtained with RL = 0.

This improvement is additionally attributed to the massive reduction in costlier TCP timeouts, in presence of link loss recovery attempts (RL = 7). It is apparent that link ARQs can improve the performance of TCP by shielding it from wireless losses as much as possible. Comparison for average cwnd with and without link loss recovery is shown in Figure 5(b). When RL = 7, TCP's average cwnd and maximum sequence number decreased linearly with increase in FER.

Figure 6(a) and Figure 6(b) represent transmission efficiency for MAC and TCP protocols with RL = 0 and RL = 7. As anticipated, link recovery (RL = 7) improves efficiency for both the layers. It must be noted that without link recovery (RL = 0), TCP's efficiency is reduced to a great extent with higher FER. Hence, link recovery improves end-to-end performance in the network by improving TCP's transmission efficiency on large scale. In Figure 6(b), about 30% of improvement for TCP's efficiency at 5% FER is recorded. Improved efficiency of MAC and TCP protocols result into significant improvement in TCP's throughput as shown in Figure 6(c).

In Figure 7(a), comparison for TCP's RTT estimated with RL = 0 and RL = 7 is presented for different FER.

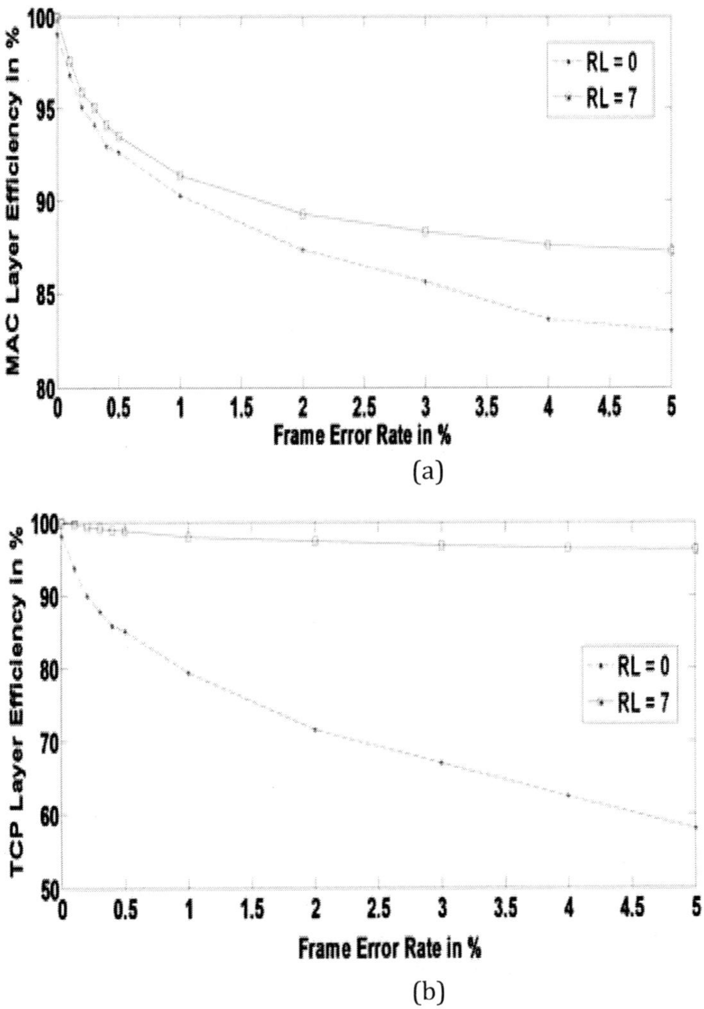

(b)

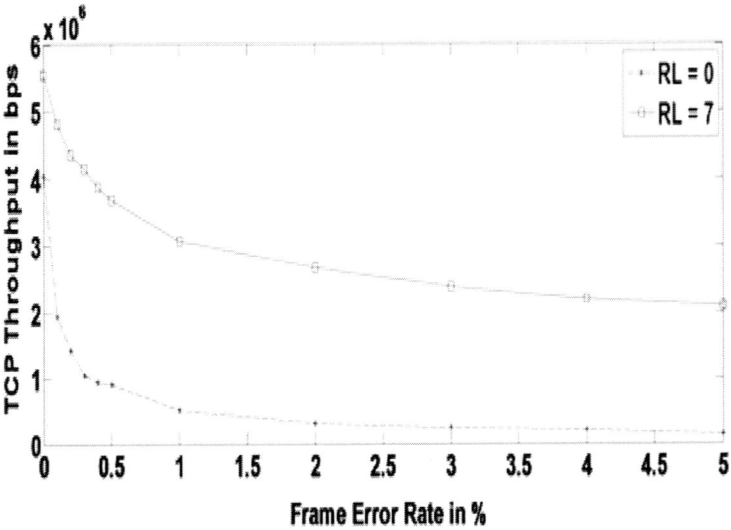

(c)

Figure 6. (a) Impact of link ARQ on MAC layer efficiency, (b) Impact of link ARQ on TCP layer efficiency, (c) Impact of link ARQ on TCP's throughput.

Without link recovery (RL = 0), the average RTT (say RTT_0) decreases with increase in FER. As shown in Figure 5(b) earlier, $cwnd_0$ has reduced sharply with increase in FER up to 2%. This reduction in $cwnd_0$ grounded for less number of TCP segments at an IFQ and hence as expected RTT_0 estimated at sender was also lowered sharply. It must be noted that the RTT_0 is found very close to the theoretical RTT in the network. A bit higher value is attributed to the contention delay at wireless interface and processing delay at nodes. In contrary, with link recovery (RL = 7), average RTT (say RTT_7) is increased along with increase in FER; as the link layer on average took a longer time for error recovery before delivering the TCP segments or acknowledgements in either directions. Figure 7(a) shows that the average RTT is increased with increase in FER until it reaches about 2%. With increase in FER above 2%, there is insignificant change in MAC retransmissions (refer Figure 3(a)) and hence similar was the impact on RTT_7 estimation. With FER above 2%, marginal reduction in RTT_7 was observed due to reduction in cwnd.

Based on the foremost observation it seems that the TCP with link layer recovery (RL = 7) utilizes network with higher value for average cwnd

($cwnd_7$) compared to that accomplished without link recovery ($cwnd_0$). Consequently, it leads to enhanced TCP's performance. In fact TCP with link recovery has achieved higher value of average cwnd at the cost of additional rise in average RTT estimation.

Since TCP's throughput is proportionate to its effective sending rate (i.e. cwnd/RTT), it is apparent that the net improvement in end-to-end TCP throughput is realized only when ($cwnd_7/RTT_7$) > ($cwnd_0/RTT_0$). In Figure 7(b), average value of $cwnd_7/RTT_7$ and $cwnd_0/RTT_0$ is plotted for different value of FER. It is revealed that the

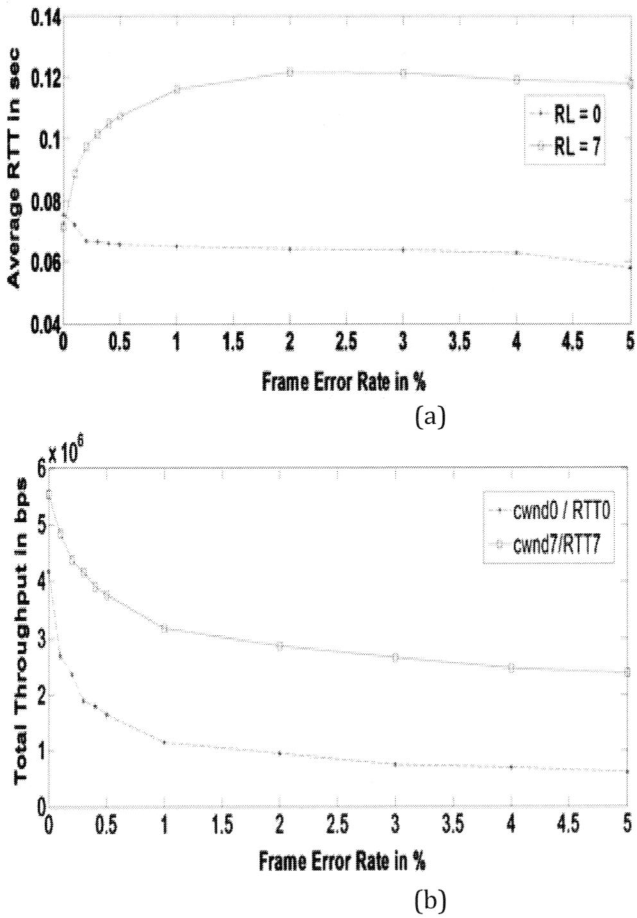

Figure 7. (a) Impact of link ARQ on TCP's RTT Estimate, (b) Impact of link ARQ on TCP's RTT Estimate.

improvement in TCP's throughput is following the net improvement for average cwnd/RTT. Pl. note that even with 0% FER (introduced using error model), transmission losses were witnessed due to channel contention (due to transmission of TCP data and ack packets on the same channel in reverse direction). Link ARQs recovered from majority of contention losses and hence higher TCP's throughput is seen compared to that observed without link recovery.

4.3. Analysis with Multiple TCP Flows

After analyzing impact of link ARQ on TCP's performance in non-congested erroneous wireless network, our simulations were extended for further investigations in a congested wireless network. For that number of simultaneous TCP senders in the networks were increased in the range of 2,4,8 and 16. The statistics obtained for FER of 0.001% are as shown in Figure 8.

As seen from Figure 8(a), with increase in number of competing flows, effective value of cwnd (for all flows over a bottleneck link) reduces with or without loss recovery as anticipated. However, for given number of competing flows, effective value of average cwnd is found much higher when link recovery was enabled. With increase in the number of TCP flows, congestion in the network increases.

Moreover link recovery in presence of transmission losses gave additional rise in RTT (as shown in Figure 8(b)). This aggravated for congestion in the network and consequently large number of congestion drops are witnessed, particularly with increase in the competing flows (for a given error rate). This has remarkably pulled down growth in cwnd due to rise in RTT, which can be seen from the Figure 8(c). As mentioned in the Figure, with link recovery (RL = 7), improvement in cwnd/RTT is seen when there was a single TCP flow. However with increase in number of TCP flows improvement for cwnd/RTT using link recovery is found insignificant. In fact when number of competing TCP flows increased upto 4 or above, link recovery failed in protecting cwnd/RTT and consequently, as presented in Figure 8(d), TCP performance is found deteriorated.

This gives insight to the issue related to the TCP's performance over an IEEE 802.11 network, even in presence of link recovery mechanism. The problem really occurs when link layer mitigation substantially diminishes growth in cwnd by increasing RTT to a much higher value with increase in FER. The similar problem may occur when link layer mitigation fails due to substantial underlying error conditions. In fact use of link layer approaches may degrade performance especially in presence of highly variable error rates. In the stated situation, reduction in cwnd and rise in RTT is unavoidable at sender on account of failed ARQs. This adversely effects on the throughput conceived by TCP.

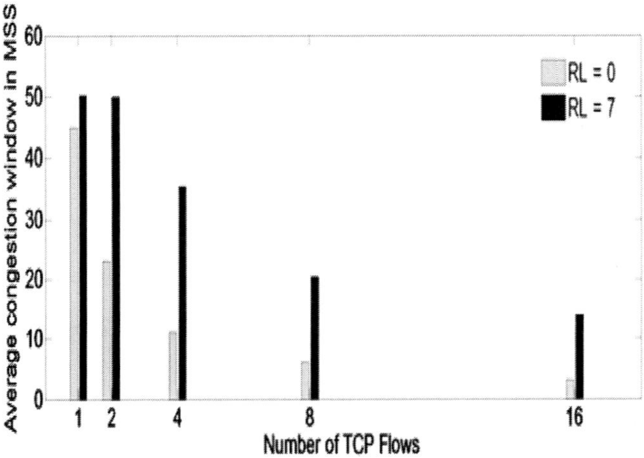

(a)

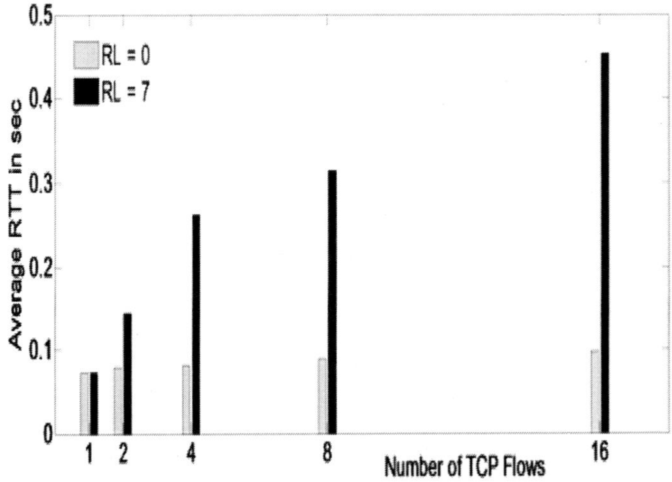

(b)

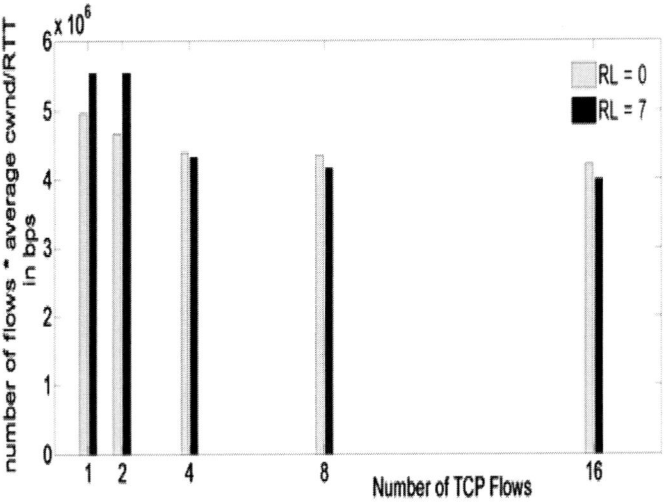

(c)

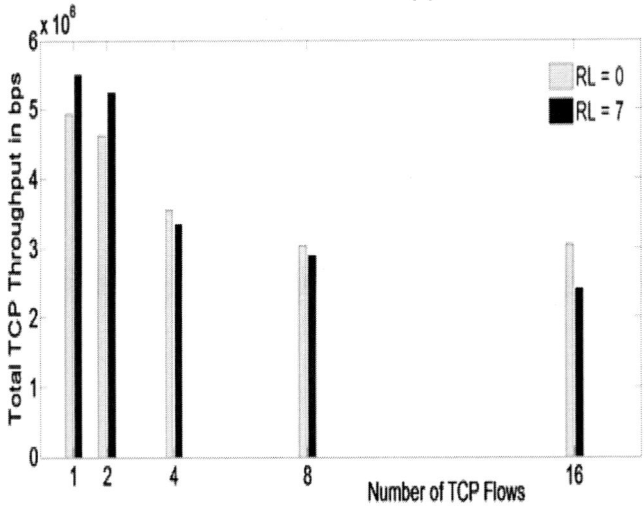

(d)

Figure 8. (a) Impact of link ARQ on average congestion window, (b) Impact of link ARQ on average RTT, (c) Impact of link ARQ on effective cwnd/RTT, (d) Impact of link ARQ on total TCP throughput.

5. CONCLUSIONS

We have performed a detailed analysis of using link ARQs to improve the end-to-end performance of TCP in wireless networks. The activities are concluded with the following outcomes.

1) Link recovery attempts using ARQ in IEEE 802.11 networks effects on two important TCP parameters; i.e. cwnd and RTT. TCP performance is decided jointly by both of the above parameters; i.e. always attributed to the improvement in the cwnd/RTT value.

2) Link ARQs in absence of network congestion and light wireless errors yield superior value of cwnd/RTT in comparison to that observed without link ARQs. This in turn results in better TCP performance.

3) Recovering from all types of wireless losses using link mitigation techniques is not always possible; particularly with high and bursty error environments. This may lead to performance degradation due to unavoidable cutback in cwnd and increase in RTT.

4) Link ARQs may degrade performance in congested networks even with very small amount of wireless errors. The performance degradation is attributed to any of the following reasons:

   - When link level error mitigation fails, reduction in cwnd at TCP sender is unavoidable. Since loss recovery attempts increase RTT and unable to protect cwnd, TCP"s end-to-end performance is compromised.

   - If multiple TCP flows are sharing common wireless channel for transmission, loss recovery attempts made for one TCP flow may affect RTT of all competing flows and therefore overall network performance is found compromised.

This shows possibility for further improvement in cwnd/RTT (and hence TCP throughput) using a corrective mechanism that makes RTT unaffected from link recovery attempts. Nevertheless, any correction that reduces RTT inappropriately may cause the sender to timeout earlier when the retransmission is being performed on the wireless link. Therefore, a well thought upon approach for RTT correction may be considered for maximizing TCP's throughput in existence of link layer ARQs. This highlights the scope for further performance enhancement of TCP proposals those advocate for retaining of cwnd based on loss discrimination.

REFERENCES

1. (2003) IEEE Standard for Information Technology-Telecommunications and Information Exchange between Systems-Local and Metropolitan Area Networks-Specific Requirements-Part 11: Wireless LAN Medium Access Control (MAC) and Physical Layer (PHY) Specifications. ANSI/IEEE Std 802.11, 1999 Edition (R2003), I-513.
2. Inamura, H., Ludwig, R. and Khafizov, F. (2003) TCP over Second (2.5G) and Third (3G) Generation Wireless Networks. IETF RFC 3481.
3. Lohier, S., Doudane, Y.G. and Pujolle, G. (2007) Cross-Layer Loss Differentiation Algorithms to Improve TCP Performance in WLANs. Springer US Journal of Telecommunication Systems, 36, 61-72.
4. Floyd, S., Henderson, T. and Gurtov, A. (2004) RFC 3782: The NewReno Modification to TCP's Fast Recovery Algorithm. IETF RFC 3782.
5. Kliazovich, D., Gerla, M. and Granelli, F. (2007) Performance Improvement in Wireless Networks using Cross Layer ARQ. The International Journal of Computer and Telecommunications Networking, 51, 4396-4411.
6. Leung, K.-C. and Li, V.O.K. (2006) Transmission Control Protocol (TCP) in Wireless Networks: Issues, Approaches and Challenges. IEEE Communications Surveys & Tutorials, 8, 64-79.
7. Sardar, B. and Saha, D. (2006) A Survey of TCP Enhancements for Last-Hop Wireless Networks. IEEE Communications Surveys & Tutorials, 8, 20-34.
8. Tian, Y., Xu, K. and Ansari, N. (2005) TCP in Wireless Environments: Problems and Solutions. IEEE Communications Magazine, 43, s27-s32.
9. Padhye, J., Firoiu, V., Towsley, D. and Kurose, J. (1998) Modeling TCP Throughput: A Simple Model and Its Empirical Validation. Proceedings of SIGCOMM, 303-314.
10. Ludwig, R. and Gurtov, A (2005) The Eifel Response Algorithm for TCP. IETF RFC 4015.
11. Barakat, C. and Altman, E. (2002) Bandwidth Tradeoff Between TCP and Link-level FEC. Computer Networks, 39, 133-150.
12. Singh, J.P., Li, Y., Bambos, N., Bahai, A., Xu, B. and Zimmermann, G. (2007) TCP Performance Dynamics and LinkLayer Adaptation Based Optimization Methods for Wireless Networks. IEEE Transactions on Wireless Communications, 6, 1864-1879.

13. Barakat, C. and Fawal, A.A. (2004) Analysis of Link-Level Hybrid FEC/ARQ-SR for Wireless Links and Long-Lived TCP Traffic. Performance Evaluation, 57, 453-476.
14. Cheng, R.S. and Lin, H.T. (2008) A Cross-Layer Design for TCP End-to-End Performance Improvement in Multi-Hop Wireless Networks. Computer Communications, 31, 3145-3152.
15. Chan, M.C. and Ramjee, R. (2004) TCP/IP Performance over 3G Wireless Links with Rate and Delay Variation. INFOCOM.
16. Luby, M., Vicisano, L., Gemmell, J., Rizzo, L., Handley, M. and Crowcroft, J. (2002) RFC 3452: Forward Error Correction (FEC) Building Block. IETF RFC 3452.
17. Kim, D., Choi, Y., Jin, S., Han, K. and Choi, S. (2008) A MAC/PHY Cross-Layer Design for Efficient ARQ Protocols. IEEE Communications Letters, 12, 909-911.
18. Scott, J. and Mapp, G. (2003) Link Layer-Based TCP Optimization for Disconnecting Networks. ACM SIGCOMM Computer Communications Review, 33, 31-42.
19. Sun, F., Li, V.O.K. and Liew, S.C. (2004) Design of SNACK Mechanism for Wireless TCP with New Snoop. IEEE Wireless Communications and Networking Conference, 5, 1046-1051.
20. Sinha, P., Nandagopal, T., Venkitaraman, N., Sivakumar, R. and Bharghavan, V. (2002) WTCP: A Reliable Transport Protocol for Wireless Wide-Area Networks. Wireless Networks, 8, 301-316.
21. Border, J., Kojo, M., Griner, J., Montenegro, G. and Shelby, Z. (2001) RFC 3135: Performance Enhancing Proxies Intended to mitigate Link Related Degradation. IETF RFC 3135.
22. Tickoo, O., Subramanian, V., Kalyanaraman, S. and Ramakrishnan, K.K. (2005) LT-TCP: End-to-End Framework to Improve TCP Performance over Networks with Lossy Channels. IEEE International Workshop on Quality of Service.
23. Vaidya, N. and Mehta, M. (2002) Delayed Duplicate Acknowledgments: A TCP-Unaware Approach to Improve Performance of TCP over Wireless. Wireless Communications and Mobile Computing, 2, 59-70.
24. Kliazovich, D. and Granelli, F. (2005) DAWL: A Delayed-ACK Scheme for MAC-Level Performance Enhancement of Wireless LANs. ACM/Kluwer Journal on Mobile Networks and Applications (MONET), 10, 607-615.

25. Kim, J., Koo, J. and Choo, H. (2007) TCP NJ+: Packet Loss Differentiated Transmission Mechanism Robust to High BER Environments. Ad Hoc and Sensor Networks, Springer Wireless Networks, Next Generation Internet Lecture Notes in Computer Science, 4479, 380-390.
26. Marcondes, C., Sanadidi, M.Y., Gerla, M. and Shimonishi, H. (2008) TCP Adaptive Westwood-Combining TCP Westwood and Adaptive Reno: A Safe Congestion Control Proposal. ICC '08, IEEE International Conference on Communications, Beijing, 19-23 May 2008, 5569-5575.

4 CHAPTER

CLASS OF SERVICE SUPPORT LAYER FOR WIRELESS MESH NETWORKS

Jukka SUHONEN, Timo D. HÄMÄLÄINEN, Marko HÄNNIKÄINEN

ABSTRACT

This paper presents an add-on Class of Service (CoS) layer for wireless mesh networks. The proposed protocol is applicable for contention-based MACs and is therefore compatible with most of the Wireless Local Area Network (WLAN) and Wireless Sensor Network (WSN) protocols. The protocol has a locally centralized control for managing data flows, which either reserve a fixed bandwidth or are weighted by fair scheduling. The protocol reduces transmission collisions, thus improving the overall throughput. IEEE 802.11 adhoc WLAN has been taken as a platform for simulations and prototyping for evaluating the protocol performance. Network Simulator Version 2 (NS2) simulations show that the CoS protocol efficiently differentiates bandwidth, supports bandwidth reservations, and reaches less than 10 ms transfer delay on IEEE 802.11b WLAN. Testing with a full prototype implementation verified the performance of the protocol.

Keywords: Wireless LAN, Wireless Sensor Network, Scheduling, Traffic Control (Communication), Quality of Service

1. INTRODUCTION

A wireless mesh network consist even thousands of devices communicating with multihop routing. Examples of wireless mesh technologies are Wireless LANs (WLANs) and Wireless Sensor Networks (WSNs). Compared to WSNs, WLAN has typically higher capacity and comprises more powerful devices such as laptops and cell phones. A WSN consists of small, low cost, and autonomic devices combining environmental sensing, data processing, and wireless communication. Unlike the traditional computer networks, a WSN is application-oriented and deployed for a specific task e.g. monitoring a physical phenomenon, object tracking, or control purposes. The WSN application space ranges from home and industrial automation to military uses. WSN nodes are typically deployed without planning in large quantities, necessitating self-configurability and distributed operation.

In general, the wireless networking offers easy deployment and mobility but has limited capacity due to constraints in radio frequency bands. Still, bandwidth is needed e.g. for file transfer and video conferencing in WLANs or the retrieval of surveillance images in multimedia WSNs [1]. As several devices share the common wireless medium, a network may congest causing rapidly increasing and varying transfer delays. The network congestion is prone to occur, when multiple users or devices are utilizing the wireless medium such as in open access WLANs or dense WSNs. Thus, Quality of Service (QoS) guarantees are required to ensure sufficient bandwidth and delays for each application.

Varying QoS requirements necessitates traffic differentiation to ensure that the important traffic is preferred when a network congests. Lower priority traffic should receive least bandwidth while higher priority traffic should get low channel access delays. Also, to prevent service degradation, a certain bandwidth should be guaranteed for constant bit rate applications such as voice and video.

The wireless medium can be shared with contention-based or contention-free approaches. The contention based approach, such as Carrier Sense Multiple Access (CSMA), has low overhead and divides bandwidth on-demand basis. However, as the offered load increases, additional methods are required to coordinate transmissions to avoid capacity loss due to collisions. Contention-free channel access eliminates collisions by assigning dedicated (reserved) communication times for a device but requires coordination between devices thus introducing additional overhead. As adjusting reservations to varying traffic requirements is complex, the majority of the popular WLANs and WSNs use CSMA based approaches or reservations are optional extensions for CSMA.

This paper presents the design, implementation, and performance evaluation of a Class of Service (CoS) support layer for wireless mesh networks referred to as Class of Service Protocol (CoSPro). CoS is an approach for delivering QoS by dividing traffic into several classes and providing differentiated service for each class. The key principle is to elect a controller device to manage traffic and grant other devices within its vicinity permissions to access medium.

Bandwidth reservation and traffic differentiation algorithms are individually well researched in the literature. However, in the existing protocols and research proposals, the coexistence of these methods requires a specific Medium Access Control (MAC) protocol. This paper presents a novel per flow QoS model that provides both bandwidth guarantees with reservations and priority based traffic differentiation. CoSPro assumes only contention-based channel access and is compatible with wide range of existing networks such as IEEE 802.11 WLAN and IEEE 802.15.4 Low Rate Wireless Personal Area Network (LR-WPAN).

The overhead of the protocol is reduced by avoiding signaling on lightly loaded network when contentionbased channel access is sufficient to guarantee acceptable performance. Thus, CoSPro achieves substantially lower delays than traditional polling protocols. Unlike traditional resource reservation protocols, unused reservations are not wasted in CoSPro as the bandwidth control algorithm assigns unused capacity to the priority based traffic flows. The feasibility of the protocol is verified with simulations and an implementation on IEEE 802.11 WLAN. IEEE 802.11 is used as an example of throughput oriented ad-hoc network technology. The results are generalizable to other contention-based MAC protocols and multimedia WSNs utilizing bandwidth critical applications.

The paper is organized as follows. Section 2 presents the related research proposals. Section 3 describes the design of CoSPro. Section 4 provides the protocol performance evaluation and simulation results. Section 5 presents the implemented prototype and gives the measurement results and Section 6 concludes the paper.

2. RELATED WORK

This section discusses the relation and applicability of CoSPro to the popular wireless WLAN and WSN standards and the related wireless QoS proposals. As CoSPro is a MAC layer enhancement, only the standards that define channel access are considered. In the related QoS proposals, proprietary QoS proposals that define completely new MAC or routing layers are not listed due to incompatibility with the existing standards.

2.1. Wireless MAC Standards

IEEE 802.11 is the most widely utilized WLAN standard that provides the nominal data rates of 11 Mbit/s, 54 Mbit/s, and 150 Mbit/s with 802.11b, 802.11a/802.11g, and 802.11n extensions. The 802.11n standard can achieve even 600 Mbit/s data rate by multiplexing spatially up to four data streams. The standard defines two topologies, infrastructure and ad-hoc. In the ad-hoc topology, stations communicate directly with each other under the Distributed Coordination Function (DCF). In the infrastructure mode, stations communicate through an Access Point (AP) under DCF or with an optional Point Coordination Function (PCF) that provides contention-free communication via reservations. A common MAC protocol based on Carrier Sense Multiple Access with Collision Avoidance (CSMA/CA) is utilized on top of the heterogeneous radio layer. The probability of collisions is reduced by carrier sensing and waiting a predetermined Inter-Frame Space (IFS) and a randomized backoff period before the transmission of a frame. Also, an optional Request to Send/Clear to Send (RTS/CTS) procedure can be used to avoid collisions due to hidden node problem [2]. However, the RTS/CTS handshake before each transmission causes additional sending delays and introduces a relatively large overhead when the size of a data payload is small.

IEEE 802.11e extension defines QoS support for 802.11 MAC by introducing two new communication modes: Enhanced Distributed Channel Access (EDCA) and Hybrid Coordination Function (HCF). EDCA is an extension to DCF, while HCF is used in the infrastructure mode. Traffic differentiation is realized by defining up to 8 traffic categories (TC) per station. A traffic category is associated with certain CW and IFS settings, in which lower values increase the likelihood for channel access and thus better throughput and lower delays. Similarly to the PCF, HCF allows stations to request resources and thus gain a reserved throughput.

The performance of IEEE 802.11e has been extensively studied. Although studies have shown that 802.11e provides relatively good service differentiation [3,4], the standard has also shortcomings. While the QoS provided by EDCA have been found notably better than besteffort, it does not have real QoS guarantees [5,6]. Also, the adjustment of CW and IFS sizes is problematic [7]. With CW, high-priority traffic susceptible to degradation when heavy low-priority exists. IFS size adjustment has been found to provide more efficient service differentiation, but it tends to starve lower priority traffic.

IEEE 802.11s is draft amendment for mesh networking. It specifies a mesh routing protocol and a new coordination function that allows reservations without a centrally coordination device. The reservation method is based on scheduled channel usage. A source device negotiates either a periodic or one-time dedicated channel access time with the destination device. During the negotiation, the destination exposes the reservation schedules of its neighbors, which allows finding a non-conflicting communication time [8]. The priority in which reservations are granted is based on the IEEE 802.11e and shares its benefits and problems.

Unlike WLANs, WSNs do not have a dominating standard as resource constrained hardware necessitates a trade-off between energy, complexity, and performance thus denying a fit-for-all protocol. IEEE 802.15.4 LRWPAN, WirelessHART, and ISA 100.11a are the three most prominent standards defining WSN MAC. In addition to these standards, the use of proprietary, deployment specific techniques is popular.

IEEE 802.15.4 uses CSMA/CA approach that is similar to the 802.11. A high-band physical layer operating in the 2.4 GHz band uses 250 kbit/s nominal data rate, while the nominal data-rates in 868 MHz and 915 MHz bands are 20 kbit/s or 40 kbit/s, respectively. IEEE 802.15.4 network supports three types of network devices: a Personal Area Network (PAN) coordinator, coordinators, and devices. The PAN coordinator initiates the network and operates often as a gateway to other networks. Coordinators collaborate with each other for data routing and network self-organization. Devices do not have data routing capability and can communicate only with coordinators. An optional low duty cycle operation allows nodes to save energy as the transceiver is active only part of the time. To reduce complexity, RTS/CTS procedure is not supported and backoff period is not increased upon collisions, which increases the probability of collisions and therefore decreases performance. IEEE 802.15.4 allows contention-free slots but does not otherwise support QoS.

WirelessHART and ISA 100.11a define a whole protocol stack including application profiles, routing and transport, and MAC. WirelessHART builds on top of the IEEE 802.15.4 physical layer but defines own MAC protocol. WirelessHART guarantees certain capacity and latency by globally scheduling transmission times for flow on each device. While the approach is acceptable for industrial applications in which reliability is the main concern and network is static and predictable, the approach is too inflexible for generic networks as it requires global knowledge of the traffic patterns. ISA 100.11a reuses IEEE 802.15.4 MAC and physical layers, but adds channel hopping and blacklisting. The network and transport layers in ISA 100.11a are IPv6 based and support various transport services, such as best-effort and real-time.

In WSNs, the scope of the CoSPro differs from Wireless HART and ISA 100.11a that are targeted at industrial, low-traffic networks. CoSPro is targeted at higher rate multimedia WSNs where network congestion is a problem.

CoSPro is compatible with IEEE 802.11 and 802.15.4 standards and solves several of their shortcomings. Differentiated traffic flows can utilize unused reservations, unlike in 802.11/802.15.4 standards where the unused reservations waste capacity. In addition, the CoSPro traffic control limits competition on air interface, and thus prevents performance drop caused by collisions typically seen in these protocols [9–11].

2.2. Wireless QoS Proposals

Several protocols and methods have been introduced for providing QoS over the Internet Protocol (IP). Popular choices are Resource Reservation Protocol (RSVP) and Differentiated Services (DiffServ) defined by the Internet Engineering Task Force (IETF). RSVP uses resource reservation, while DiffServ offers CoS by utilizing the IP header Type of Service (TOS) field and defining a basic set of rules for differentiating packet forwarding in the routers. DiffServ and RSVP can be used in conjunction of other protocols. For example, the IETF defined Multi-Protocol Label Switching (MPLS) has a support for DiffServ. MPLS makes network management easier for QoS by creating a specific path for a sequence of packets. Also, Subnet Bandwidth Manager (SBM) is a RSVP based admission control protocol targeted for IEEE 802-style LANs. The IP-based methods are complementary to the CoSPro, as they ensure end-to-end QoS and can utilize CoSPro to manage per-hop service.

Several modifications for providing CoS for WLAN have been proposed in research papers. Most of these concentrate on improving the WLAN MAC protocol. The proposed modifications often utilize the changing of the size of the contention window or backoff interval of the CSMA-algorithm [12,13]. Another method is to change the length of the IFS. Since a shorter IFS gives a higher priority access, some proposals present an enhanced service support by changing the type or size of IFS [14,15]. While the concepts of these methods can be used with both DCF and PCF, few modifications concentrate solely on PCF [16,17]. As the IEEE 802.15.4 channel access is conceptually similar to the 802.11 WLAN, similar backoff and IFS based QoS have been proposed [10,18,19]. Other related methods for achieving CoS on wireless networks include reservation based protocols [20], enhanced support and adaptation of a particular higher layer protocol (mainly UDP and TCP) [21,22], and admission control [23].

The protocol presented in [23] estimates available bandwidth and prevents congestion with admission control for real-time flows and rate control for non-real time flows. In the protocol, the bandwidth usage estimation is a critical issue and the protocol needs to be integrated with MAC to get the extensive channel usage information.

A number of research articles address the performance problems encountered in the IEEE 802.11 and 802.15.4 channel access. Performance can be enhanced by adapting to the channel in order to compensate channel errors [24,25], by tuning the CSMA algorithm to improve throughput [26], or by avoiding hidden node collisions in multihop networks [27]. However, these MAC layer proposals do not usually provide differentiated QoS.

A problem with the MAC layer research proposals for QoS is the guaranteeing of a certain bandwidth for a single flow. The bandwidth reservation can be provided by a network layer solution, but all traffic belonging to the bandwidth reserved with such method must still contend with other traffic on the link layer. The limited capacity and varying radio environment in wireless networks demand constraining offered load [28,29].

Unlike most of the related proposals, CoSPro provides both bandwidth control according to the application requirements and solves the performance degradation seen in IEEE 802.11 and 802.15.4 with a high number of devices. CoSPro can be implemented without modifying the underlying MAC layer. Also, the bandwidth control supports both bandwidth reservation and traffic classification.

3. COSPRO DESIGN

The lack of QoS is generally not a problem if network is lightly loaded, as all devices get desired bandwidth and buffering delays remain small. The design of CoSPro relies on this fact by controlling bandwidth usage to prevent network congestion and thus performance problems. Minimizing control overhead is another key factor in the design; control messaging is relaxed when the network is not congested.

CoSPro uses locally centralized approach to control traffic in one-hop radius. The design extends to mesh networks, where separate parts of the network are controlled independently. In mesh networks, nodes make forwarding decisions independently resulting into a selfconfiguring and self-healing multi-hop topology. With CoSPro, each mesh node belongs to a certain local traffic control area managed by a controller device. The controlled devices are referred to as end devices. Data transfer can take place between the controller and end devices or directly between end devices. In mesh networks, a controller manages traffic in a virtual cluster of devices as shown in Figure 1. The clustering algorithm should minimize the number of clusters by selecting one controller within communication range. Suboptimal clustering (several or no controllers for an end device) lowers performance but does not prevent device operation, as CoSPro is built on top of a contention-based channel access that will work regardless of the existence of a CoSPro controller. Defining the clustering algorithm is outside the scope of this paper, but examples of suitable algorithms are presented in [30] and [31]. CoSPro can be applied directly to clustered networks, such as in IEEE 802.15.4 operating in cluster-tree mode. In these networks, cluster heads forward traffic from their child nodes. This way, a child node saves energy as its transceiver can be turned off most of the time.

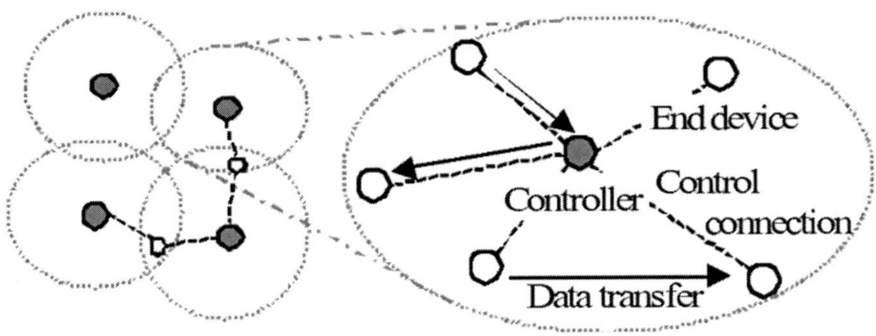

Figure 1. The CoSPro mesh network topology.

CoSPro controls traffic per hop basis and can thus be realized either on the mesh routing or MAC layer as shown in Figure 2. The implementation on MAC layer decreases overhead as messages and status information can be piggybacked to existing Protocol Data Units (PDUs), but may not be feasible e.g. when MAC protocol is hardware implemented. An application can utilize CoSPro directly through an Application Programming Interface (API), in which case application packets are fitted into a CoSPro PDU, or indirectly by differentiating traffic based on its contents (e.g. TOS field). In the latter case, traffic is controlled by predefined rule sets and CoSPro is transparent to applications, allowing e.g. the use of IP protocols with the IPv6 over Low power WPANs (6LoWPAN) technique. With 6LoWPAN, CoSPro is implemented on mesh routing layer (mesh under) instead of IP layer (route over). The protocol analysis and implementation in this paper assume the routing layer and the API approaches.

The functional model of the CoSPro end device is presented in Figure 3. The CoSPro classifier prepares application data for sending and puts data to transmission queue according to its traffic class. The local scheduler retrieves packets from queue according to the flow settings. The phases in sending of a PDU are managed by the CoSPro control, which communicates with the controller device. The CoSPro control also allows an application to set flow priorities and ask for bandwidth reservations. The received data is passed transparently to the application.

The model of the controller device is presented in Figure 4. The network scheduler is used for handling differentiated traffic and the reservation table holds reserved flow information. CoSPro uses the medium adaptation to provide information about the usage of the medium and its capabilities, such as data rate. The usage of the medium affects the decision to enter or leave the congestion mode, whereas reservations utilize knowledge of the available data rates. A configuration for limiting allocations and the value of the congestion threshold is determined by the policy control.

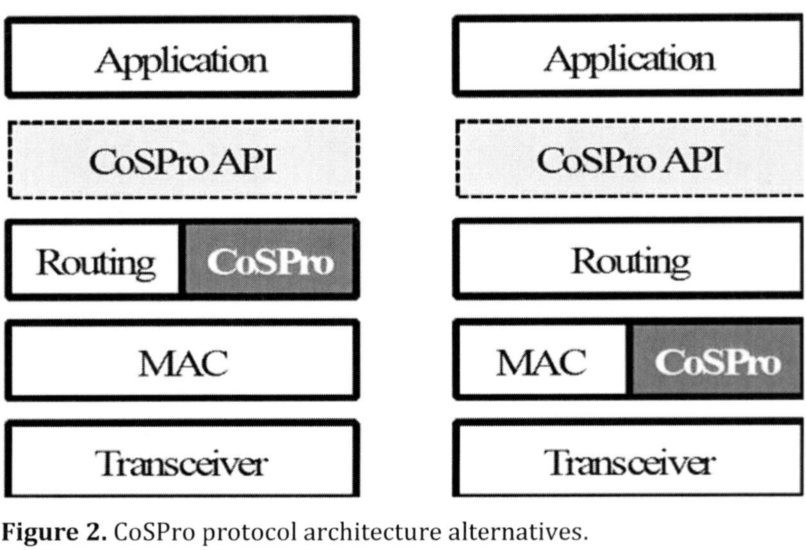

Figure 2. CoSPro protocol architecture alternatives.

3.1. Traffic Management

The goal of the CoSPro traffic management is a versatile traffic differentiation while assuming only simplest of contention based channel access protocols. Both prioritized channel access and bandwidth guarantees are supported, and applications using different types of traffic can coexist seamlessly in a network.

CoSPro manages traffic in data flows. Each end device can originate or terminate one or more flows. A flow uses one of the two modes provided by CoSPro, which are the reserved bandwidth mode and differentiated bandwidth mode. The modes define how the available bandwidth is divided between active devices. Both reserved and differentiated mode flows coexist in the network. The reserved bandwidth mode allocates a fixed bandwidth for a flow and is thus suitable for satisfying flows with strict bandwidth requirements. CoSPro uses the remaining capacity for the differentiated bandwidth mode traffic, which provides controllable bandwidth division according to traffic classes. A controller manages flows only in one-hop radius. For multi-hop reservations, CoSPro can be used with conjunction of end-to-end bandwidth reservations protocols such as RSVP.

The CoSPro controller keeps record of average bandwidth usage in the network by receiving periodic channel usage updates from active devices. A device that has not transmitted traffic does not need to send updates. An end device can send freely with the differentiated mode as long as the used bandwidth remains under a network specific congestion threshold. When the threshold is exceeded, CoSPro considers the network to be congested and the devices must request a permission to send differentiated mode flows. The use of congestion threshold significantly reduces signaling overhead on lightly loaded network, since there is no need to exchange request and grant signals.

The QoS states used in a CoSPro end device is depicted in Figure 5. An end device changes its congestion state upon receiving an indication from the controller. All flows in a device share the congestion state. A flow is initially in the differentiated bandwidth mode, but enters the reserved bandwidth mode when the end device containing the flow requests and receives a reservation update from the controller device. The flow is set back to differentiated mode, if the reservation is dropped.

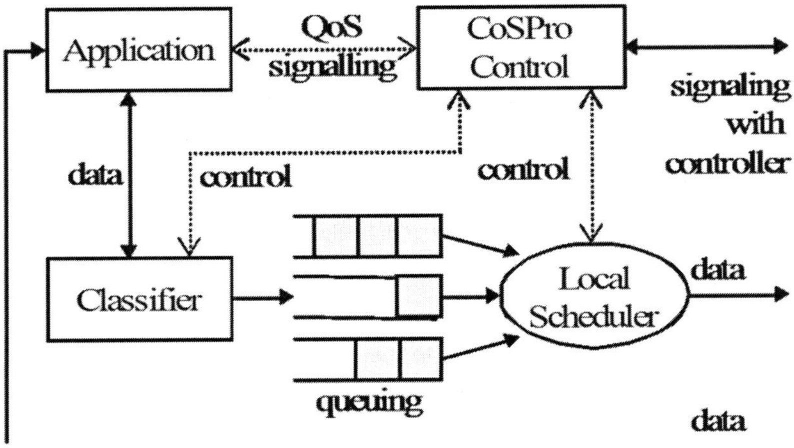

Figure 3. Functional architecture of the CoSPro end device.

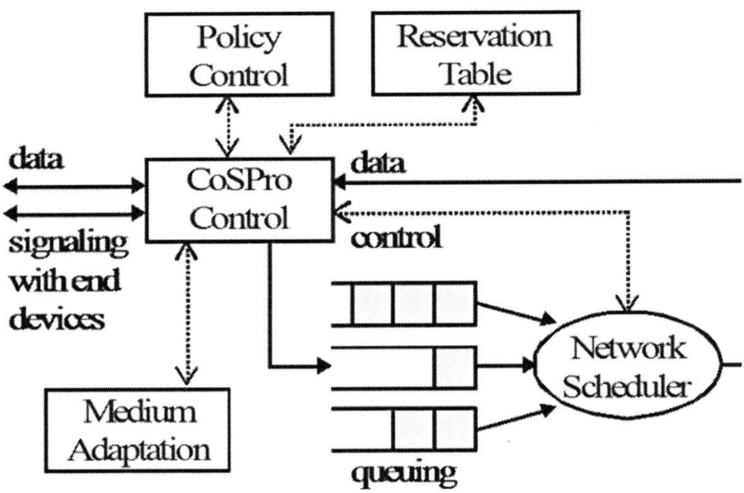

Figure 4. Functional architecture of the CoSPro controller.

3.2. Traffic Classes

Traffic class is a mean to define and provide support for different traffic types. For ease of configurability, a traffic class must be defined in an understandable manner. For example, the level of service in 802.11e defined traffic classes (background, best-effort, video, and voice) related to each other is not obvious. Adjusting the traffic classes for the exact level of service requires experimenting with different values. The definition of CoSPro traffic class allows predictable and easily configurable level of service: bandwidth is divided fairly based on defined weights (priorities).

A CoSPro traffic class is consists of priority and aging time parameters. Each flow is assigned with a traffic class that can be negotiated between controller and an end device. For example, in the prototype implementation, an end device defines the used traffic class by sending the parameters during an initial handshake procedure with the controller. The aging time defines the maximum transfer time for a PDU. After the aging time has elapsed, a queued packet is discarded as obsolete. The interpretation of the priority parameter differs slightly between the reserved and differentiated bandwidth modes. In the reserved traffic mode, priority indicates whether the controller device should accept or reject a new allocation request. In the differentiated mode, the priority parameter defines how the bandwidth is di vided among traffic flows. Also, low priority reservations can be dropped if differentiated traffic contains a lot of high priority flows.

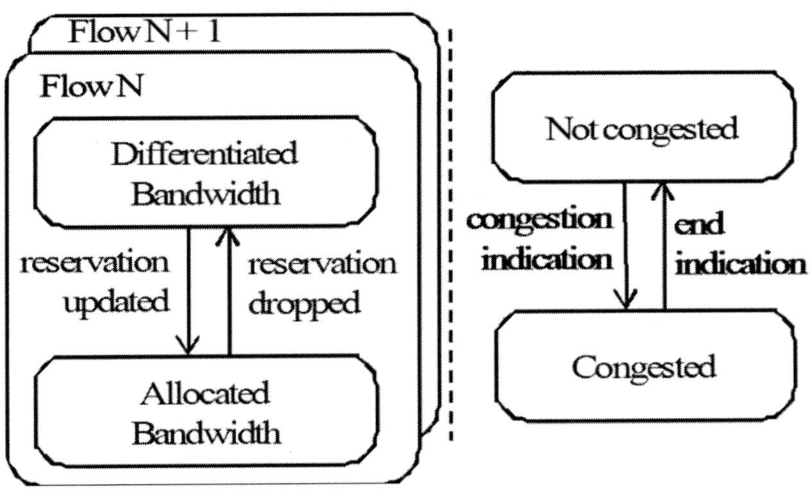

Figure 5. QoS states in a CoSPro end device.

3.3. Reserved Bandwidth Algorithm

In the reserved bandwidth mode, controller grants a portion of total capacity for a flow. The reserved bandwidth mode relies on the fact that contention based MAC protocols support strict QoS requirements, when offered traffic load is controlled and not near the maximum capacity [32]. Therefore, by ensuring that used bandwidth does not exceed the maximum capacity, collisions are rare and each flow gets statistical throughput and delay guarantees.

A device reserves bandwidth by sending a request containing minimum required and preferred bandwidth to the controller. Controller assigns each reserved flow at least the minimum requested bandwidth. If the requested bandwidth exceeds available capacity, lowest priority flows are dropped. Surplus bandwidth is assigned first to the higher priority flows. Controller may redefine or cancel the reservation to enable admission for a higher priority flow. The local scheduler of an end device limits traffic to the negotiated bandwidth by delaying the retrieval of next packet for a flow, thus limiting the number of reserved flows accessing air interface simultaneously.

3.4. Differentiated Bandwidth Algorithm

In the differentiated bandwidth mode, an end device requests the controller for permission to send data. This is a common approach in polling mechanisms that has the main drawback of increased overhead and latency [33]. CoSPro introduces several methods to improve the polling efficiency. First, polling is used only when the network is congested. Second, a node may send several packets when it gets the permission, which reduces the per packet overhead. Third, instead of assigning each device a transmission opportunity periodically [34], CoSPro grants transmission periods dynamically based on the knowledge of flow activity and queued frames. Therefore, CoSPro is able to control latency and bandwidth.

The differentiated bandwidth mode uses a weighted fairness algorithm, in which the flows are given bandwidth in proportion to their weight. In CoSPro, a priority parameter p_i is used as the weight of a flow f_i. A flow f_i gets an average bandwidth of b_i as

$$b_i = \frac{p_i}{\sum_{j=0}^{N-1} p_j} B_{diff} \quad , (1)$$

where N is the amount of active flows and B_{diff} is the bandwidth available to differentiated mode flows. The adjustment of priority is intuitive, for example a class with priority of four receives twice as much bandwidth as class with priority of two. The value for B_{diff} is calculated as

$$B_{diff} = B_{total} - \sum_{i=0}^{N-1} r_i \quad , (2)$$

where B_{total} is the total available bandwidth and r_i is the actual bandwidth utilized by reserved mode flow f_i.

A device signals the status of its local queue within a transmission request containing information about each active flow. The controller queues the requests, uses the network scheduler to determine the next request, and sends an allowed transmit indication to an appropriate device. The indication contains the allowed traffic class, the length of the transmission period, and a limit to the sending bandwidth. The bandwidth limit is determined by (2). Thus, the limit ensures that enough capacity is available for reserved flows, but also utilizes bandwidth that is unused by reserved flows. Consequently, a differentiated mode flow can send data without bandwidth limitations when reservations do not exist.

A device stops sending, when the transmission period length is reached. Next, the device sends end transmission indication to the controller, which contains the same information than traffic request. The message denotes that the device has completed sending during this period and eliminates the need to update the flow status with a new traffic request message. The controller may interrupt an active device with deny transmit indication, if the controller has a request that must be scheduled immediately. If the controller does not receive end indication, it schedules next request after the transmission period has ended.

3.5. Differentiated Mode Scheduling Algorithm

This section presents one possible algorithm for providing bandwidth differentiation as shown in (1). The algorithm realizes bandwidth allocation by scheduling transmission opportunities based on the flow priority values, requests sent by the devices, and the amount of successfully transmitted traffic. The use of transmitted traffic ensures that bandwidth differentiation is fair between flows using different packet sizes and devices having differrent packet error rates. For example, a device experiencing high packet error rate can still get more bandwidth than lower priority flows from devices having a better link. The use of an algorithm that considers aging time as a scheduling deadline could improve real-time scheduling for traffic with same priorities [35,36]. However, it would not provide differences in bandwidths, which is the approach taken in CoSPro.

The scheduler in the CoSPro controller device has a list of active flows F. A flow f is marked active after a traffic request concerning it is received. Each request contains the number of PDUs waiting for sending at the end device, denoted as c, and the average size of waiting PDUs, denoted as a. The scheduler selects one of the active flows and gives it a permission to send PDUs for the duration of d. A flow is marked inactive after it is scheduled.

For the scheduling decision, the scheduler keeps a record of received traffic from each flow, denoted as t_i. The scheduled flow is selected by weighting the amount of received traffic by the priority of the flow. The weighted traffic value w_i is derived as

$$w_i = t_i / p_i. \tag{3}$$

The flow f_i that has the smallest weighted traffic value w_i is selected. If there are several flows having the same weight, the one with the highest priority is selected. The purpose of the selection is to keep the traffic weights equal. As the weights are equal, the transferred traffic and therefore the obtained bandwidth are proportional to the priority values as (1) requires. For ensuring that the w_i calculation is not distorted by different device activities, the scheduler reduces the recorded received traffic counters (t_i) periodically. Otherwise, the devices that have not sent for a long time would get more bandwidth than defined.

After the flow has been selected, the scheduler determines the time the device is allowed to send PDUs belonging to the flow. This is done by calculating the length of the transmission period d. First, the amount of PDUs is determined, denoted as n. Then, n is converted to a particular transmission period length d.

If F contains only one flow, the n is set directly to the value c_i contained in the request. Otherwise, the n is calculated as

$$n = \min\left(c_i, \frac{(w_k - w_i) p_i}{a_i} \right), \tag{4}$$

where w_k corresponds to flow f_k, which is the flow with the second smallest weighted traffic value in F. By comparing the two smallest weighted traffic values, the scheduler can allocate the length of the transmission period for the selected flow before the other flows (from which the traffic requests has been received at the moment) need to be scheduled for transmission. In (4), the actual number of bytes to be sent is derived by multiplying the subtracted weighted traffic values with priority pi. The amount of bytes is converted to the number of PDUs by dividing it with the average PDU size a_i. Taking the minimum prevents any larger n value than the device has requested.

Finally, the length of the transmission period is calculated from the PDU count with the average PDU size and the known network bandwidth b as

$$d = na_i \;/\; b. \tag{5}$$

The scheduler has two configurable parameters for controlling the average sending durations: D_{min} and D_{max}. The d is assigned within these limits. The limits affect to the overall CoSPro trade-off between transfer delay and throughput. The purpose of the D_{min} parameter is to improve throughput by ensuring that several PDUs can be sent during a transmission period. The D_{max} is used to control the delay that a flow experiences while waiting for a transmission period to be reserved.

4. PERFORMANCE ANALYSIS

The performance is analyzed when CoSPro is implemented on top of the IEEE 802.11 MAC operating under DCF. The DCF mode was chosen because the proposed centralized control architecture of CoSPro would be interfered by PCF.

4.1. Throughput Analysis

The throughput of CoSPro using differentiated mode is compared to the standard 802.11 with and without the RTS/CTS procedure. CoSPro reserved mode does not include additional messaging and has throughput similar to the standard WLAN. The values are calculated by assuming IEEE 802.11b having 11 Mb/s basic data rate with direct sequence spread spectrum. The overhead caused by IFS sizes, backoff times, and acknowledgements for WLAN MAC PDUs have been accounted in the calculations. The calculated bandwidth presents the best-case throughput for a single device, since the effect of collisions, bit-errors, and retransmissions is ignored. CoSPro overhead includes also transmission requests, allow transmit, and the end of transmission indications.

The calculated throughputs are shown in Figure 6. The frame overhead and acknowledgements cause considerable throughput penalty with frames having a small payload. The throughput penalty of CoSPro differentiated flows depends on the length of the transmission period. Differentiated mode flows shows better results than the standard WLAN with RTS/CTS, when the transmission period length exceeds 10 ms. Compared to the standard WLAN with 1500 B payload, CoSPro has 4.5% lower throughput with 50 ms transmission period when RTS/CTS is not used and 17% higher throughput when RTS/CTS is used. In addition, CoSPro has a good throughput with a small frame payload, because it allows more frames to be sent within a single transmission period. As the transmission period length is increased, the throughput of CoSPro approaches the standard WLAN values. Because the length of the transmission period is controlled with D_{min} and D_{max} scheduler parameters, the overhead penalty can be adjusted to an acceptable range. In general, longer transmission periods reduce the overhead.

4.2. Delay Analysis

The delivery delay for a PDU contains the time required to transfer it between MAC layer entities and the time it waits for scheduling. The delay can be evaluated by calculating the time an arbitrary flow k must wait for scheduling. Let t_k denote the amount of data belonging to the kth flow. Before sending the kth flow again, the device must wait until equal amount of data has been transferred via other flows. The amount of data is determined by the priority values and can be obtained from (1) and (3) as

$$t_i = (p_i / p_k)t_k \quad , (6)$$

where t_i is the traffic that must be transferred via any other ith flow before the device can sent to the kth flow again. The value of t_i can be converted to time d_i as

$$d_i = (p_i / p_k)d_k \quad , (7)$$

where d_k denotes the time taken to send traffic t_k. It is easily seen that a long transmission period for the kth flow increases the transmission periods of the other flows as well. Since a flow must wait for other flows to finish, the use of long transmission periods increases delays. Also, it should be noted that (7) does not include the time caused by the MAC layer overhead or the CoSPro signaling. For equal sized PDUs, the delay will be divided in the fraction of the flow priorities similarly to the bandwidth.

4.3. Simulated Performance

CoSPro performance was verified with the Network Simulator (NS) version 2 [37] and compared to IEEE 802.11 EDCA [38]. CoSPro was implemented in NS as a new protocol layer, which was located on top of the simulated 802.11 MAC layer. The simulation settings used 11 Mb/s physical link rate and did not include the generation of packet losses, RTS/CTS signaling, or fragmentation.

The simulation measured traffic differentiation and the effect of competition in air interface, by increasing the number of active devices. Each device utilized three flows with real-time, background, and best-effort priorities. Real-time flow was presented typical voice traffic, while

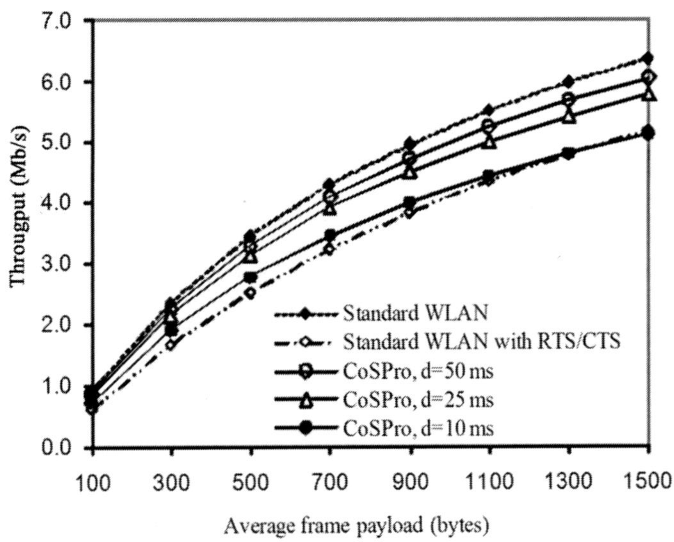

Figure 6. The effect of payload size and transmission period d in differentiated mode CoSPro.

Table 1. Simulation settings with three different traffic flows.

Flow	Packet size (B)	Offered load (kbit/s)	CoSPro Traffic mode	CoSPro Priority	802.11 EDCA AC
Background	1400	400	Differentiated	1	Background
Best-effort	800	200	Differentiated	2	Best-effort
Real-time	204	64	Reserved	-	Video

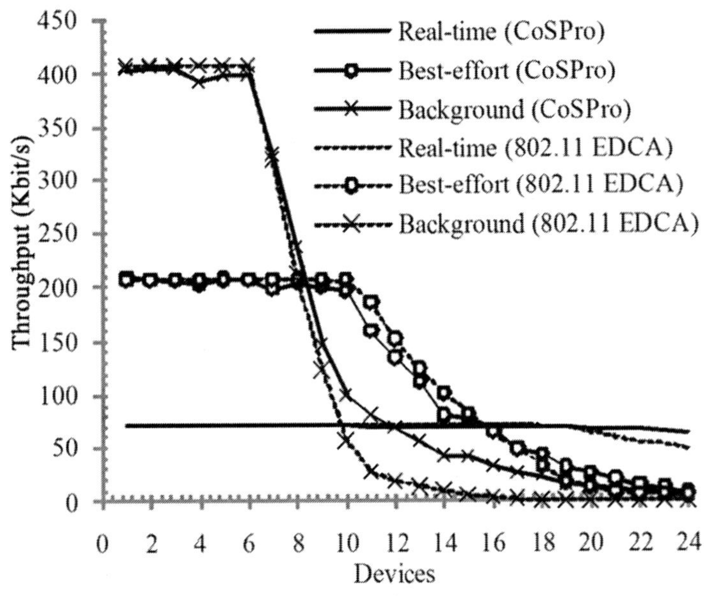

Figure 7. Average throughput of a traffic flow with CoSPro and IEEE 802.11 EDCA.

background and best-effort flows presented typical data traffic. The offered throughput and simulation settings with CoSPro and IEEE 802.11 EDCA are presented in Table 1. The 802.11 EDCA access categories determine IFS and backoff values and are the same as defined in IEEE 802.11e standard. Video access category was used for real-time traffic, because voice access category performed poorly in EDCA when the number of devices was increased.

Figure 7 presents the average throughput of traffic flows. After 6 active devices, network capacity is exceeded and the throughputs of low priority flows decrease. CoSPro provides better throughput for real-time traffic when the number of active devices increases. The performance drop with 802.11 EDCA is caused by the increased number of collisions, whereas CoSPro limits the number of sending devices thus reducing the competition. In 802.11 EDCA, higher priority traffic causes starvation to the lower priority traffic. CoSPro does not have this problem. Instead, the aggregate throughputs of best-effort and background traffic in CoSPro converge with 24 active devices because of used minimum scheduler time D_{min}, which prevents low priority traffic from starving. As the competition over air interface increases, 802.11 EDCA real-time traffic drops, while CoSPro is able to provide real-time traffic flow sufficient throughput without notable performance drop.

End-to-end packet delays with CoSPro and 802.11 EDCA with 6 active devices are shown in Figure 8. 802.11 EDCA real-time traffic has the smallest delays due to its small IFS and backoff values. As the CoSPro uses the legacy 802.11 IFS and backoff values that are

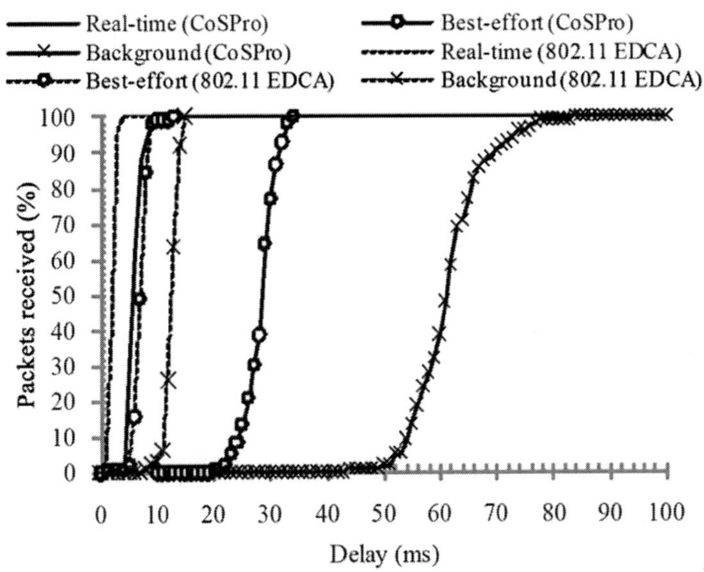

Figure 8. End-to-end packet delay in CoSPro and IEEE 802.11 EDCA with 6 active devices.

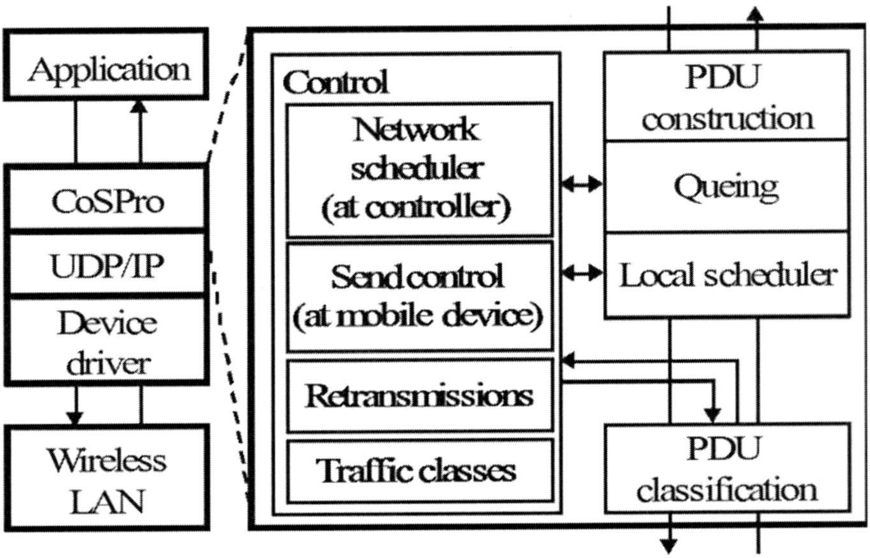

Figure 9. Prototype implementation architecture of CoSPro.

slightly higher than EDCA real-time values, the CoSPro real-time traffic has slightly higher delays. However, the delays are still low enough for the flows to be used for multimedia traffic.

5. PROTOTYPE IMPLEMENTATION

The prototype implementation places CoSPro between the application and WLAN link protocol layers, as presented in Figure 9. The prototype consists of the controller and end device software modules that implement CoSPro functionality. In the implemented prototype, CoSPro also operates on top of the User Datagram Protocol/Internet Protocol (UDP/IP) layers. The prototype was built on that layer in order to enable testing with different existing network cards and drivers. Although UDP/IP increases overhead, the layers are otherwise transparent and do not affect the operation of the prototype, since the layers do not employ flow control or acknowledgements.

The prototype architecture is the same for both the end device and the controller device except for the control part. An application data packet is fitted inside CoSPro PDU in the PDU construction function. A CoSPro PDU contains the identification of the used traffic class and the information needed to separate CoSPro control and application data in the PDU classification part of the protocol. The PDU is queued until the local scheduler selects it and sends it by using the lower protocol layers. The local scheduler is controlled by the send control part of CoSPro. The send control communicates in the congestion mode with the network scheduler, which is located at the controller device. The protocol control provides retransmissions for the CoSPro control messages.

The CoSPro modules were implemented as Microsoft Windows 32-bit executables. The executable programs utilized Windows Sockets 2 application programming interface (WinSock2) for networking. The prototype used common PC hardware with Avaya Wireless PC Card, Nokia C110/C111 and Nokia D211 WLAN cards meeting the IEEE 802.11b standard.

6. PROTOTYPE MEASUREMENTS

The measurements were carried out using the implemented prototype. The end devices were located around a central device, which was also the CoSPro controller device. The distances between the devices were less than ten meters, as the purpose was to provide similar link conditions to all devices and to minimize the near-far effect. The central device was used to receive all data sent by the other devices. The results obtained with CoSPro were compared to the standard WLAN. In the standard WLAN tests, devices sent data also to the central device, but the CoSPro software modules were not used.

The measurements utilized the 11 Mbit/s nominal bandwidth of IEEE 802.11b. Each measurement used the packet size of 1400 B, which was selected to minimize the header overhead and to avoid IP datagram fragmentation. In the used WLAN cards, the RTS/CTS procedure and fragmentation were disabled, and the number of transmission retry limit was set to four. Also, the default values were used for other MAC parameters.

6.1. Performance of Priority Scheduling

The priority scheduling test evaluated the performance of the CoSPro scheduling algorithm. The test used four devices utilizing priority values of

2, 4, 6, and 8. Each device started sending at 700 kbit/s and increased the offered throughput to 2.7 Mbit/s during the test.

Figure 10 presents the measured throughput with CoSPro. The CoSPro congestion threshold is exceeded when a device sends more than 1.0 Mbit/s, which totals to more than 4 Mbit/s network load. As a result of the protocol, the devices with higher priorities get better throughput with the expense of lower priorities. The device having the lowest priority traffic carries the penalty of additional overhead, while the other devices are more or less unaffected by the overhead. Throughput of the device using the highest priority increases, until the sending bandwidth reaches the point where the priority settings do not allow the device to increase its share of bandwidth.

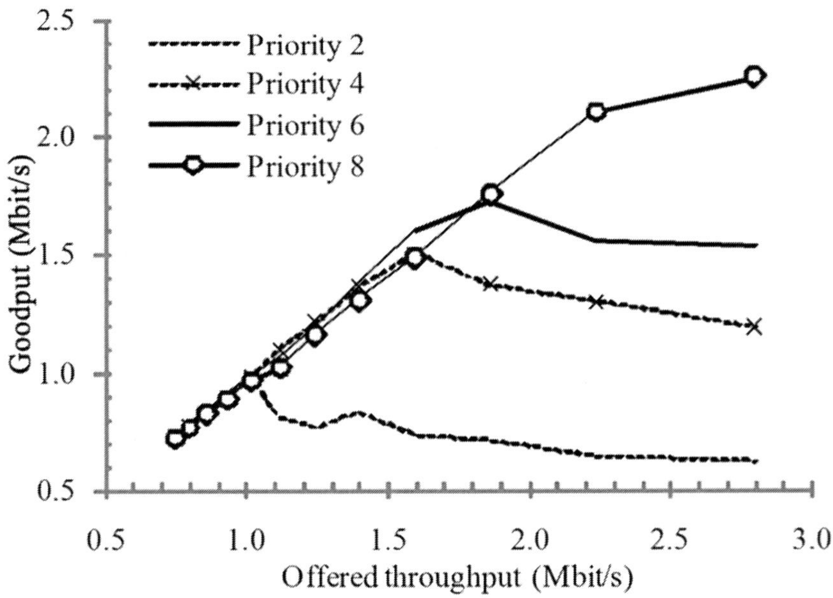

Figure 10. Device throughput with CoSPro.

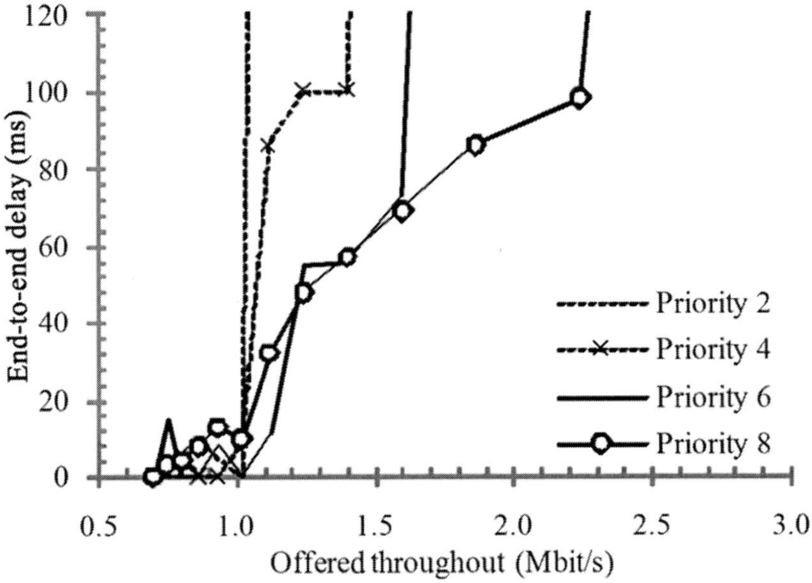

Figure 11. Packet transfer delays with CoSPro.

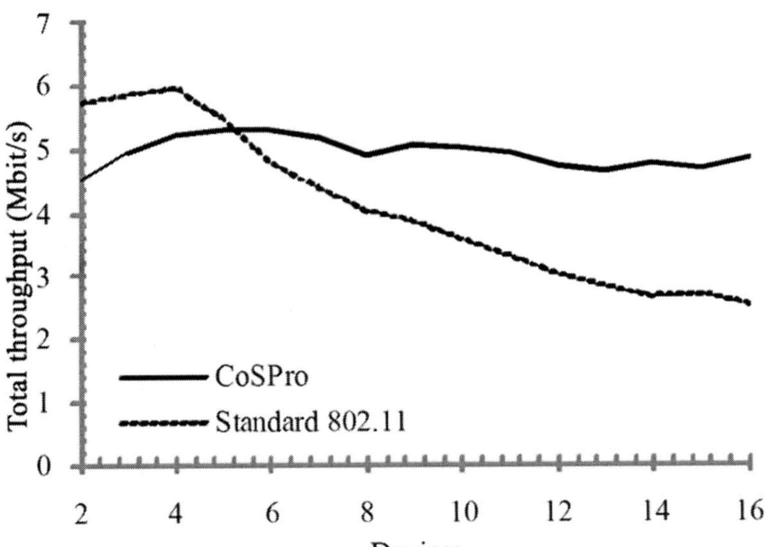

Figure 12. Measured network throughput with and without CoSPro.

The test setup used a scheduler configuration that preferred higher throughput and considered 50-100 ms delays acceptable (D_{min} and D_{max}). Therefore, while the test results show that CoSPro did not cause a notable penalty on throughput, it adds a small delay. The packet queuing and transfer delays from an end device to the controller are presented in Figure 11. End-to-end delays rise when the network congests and packets have to wait for sending. The traffic with a high priority endures well against network congestion.

6.2. Throughput in a Congested Network

The effect of an increased traffic load and congestion on the total network throughput was tested by increasing the number of active devices, while each device offered 5.0 Mbit/s throughputs. Thus, the offered throughput exceeds the IEEE 802.11b practical bandwidth limit already with two active devices.

The measured total network throughput is presented in Figure 12. The network throughput in the standard WLAN achieves its peak with four active devices. After the peak, throughput quickly drops as the number of collisions on the link increases. The signaling overhead of CoSPro is evident with small number of active devices. However, when the number of devices increases, CoSPro limits competition in air interface and provides a higher throughput than standard WLAN. In addition to outperforming the standard WLAN, the throughput stays steady as the number of devices increases.

7. CONCLUSIONS

This paper presented a CoS support layer for wireless mesh networks referred to as CoSPro. The proposed protocol is applicable for contention-based MACs and is therefore compatible with most of the popular WLAN and WSN protocols. The proposed protocol is simple but efficient, which makes it easy to implement in existing networks. CoS is realized by managing traffic flow activity with a controller device. Both the simulations and the measurements with the implemented prototype confirm that the protocol differentiates bandwidth and transfer delays for each traffic class. The protocol adds 5-20% overhead depending on configured latency/throughput trade-off, but increases overall throughput on congested network by avoiding collisions. In a non-congested network, CoSPro does not cause any overhead, but the traffic control is activated only when offered load exceeds a defined threshold. The total throughput of CoSPro exceeds the standard 802.11b WLAN with more than 6 competing devices.

The performance of CoSPro can be increased by integrating it more tightly with wireless MACs. The traffic request and response signaling can be implemented in the RTS/CTS procedure. Also, in synchronized MACs, the beacon frame can be utilized to signal the congestion state indication thus reducing CoSPro signaling overhead. On the other hand, when the protocol is implemented on higher layer, it is compatible and complementary to the wireless standards, such as IEEE 802.11 and IEEE 802.15.4.

REFERENCES

1. I. F. Akyildiz, T. Melodia, K. R. Chowdury, "Wireless multimedia sensor networks: a survey," IEEE Wireless Communications, pp. 32–39, December 2007.
2. G. Bianchi, "Performance Analysis of the IEEE 802.11 Distributed Coordination Function," IEEE Journal on Selected Areas in Communications, vol. 18, pp. 535–547, March 2000.
3. Y. Xiao, "Enhanced DCF of IEEE 802.11e to support QoS," in Proceedings of Wireless Communications and Networking Conference (WCNC), 2003, Vol. 2, pp. 1291–1296, March 2003.
4. P. Garg et al., "Using IEEE 802.11e MAC for QoS over Wireless," in Proc. Performance, Computing, and Communications Conference, pp. 537–542, 2003.
5. H. Zhu et al. "A survey of quality of service in IEEE 802.11 networks," IEEE Wireless Communication, August 2004.
6. D. He and C. Q. Shen, "Simulation study of IEEE 802.11e EDCA," in Proceedings of VTC, 2003.
7. J. W. Robinson and T. S. Randhawa, "Saturation throughput analysis of IEEE 802.11e enhanced distributed coordination function," IEEE JSAC, Vol. 22, No. 5, June 2004.
8. G. Hiertz, S. Max, and T. Junge, "IEEE 802.11s – mesh deterministic access," in Proceedings of the 14th European Wireless Conference, pp. 1–8, 2008.

9. M. Hännikäinen, M. Niemi, and T. Hämäläinen, "Performance of the Ad-hoc IEEE 802.11b wireless LAN," in Proceedings of International Conference on Telecommunications, pp. 938–945, June 2002.

10. S. Pollin et al. "Performance analysis of slotted carrier sense IEEE 802.15.4 medium access layer," IEEE Transactions on Wireless Communications, Vol. 7, No. 9, pp. 3359–3371, September 2008.

11. T-J Lee, H. R. Lee, and M. Y. Chung, "MAC throughput limits of slotted CSMA/CA in IEEE 802.15.4 WPAN," IEEE Communication Letters, Vol. 10, No. 7, pp. 561–563, July 2006.

12. A. Dugar, N. Vaidya, and P. Bahl, "Priority and fair scheduling in a wireless LAN," in Proceedings Military Communications Conference, Vol. 2, pp. 993–997, 2001.

13. D. Qiao and K. G. Shin, "Achieving efficient channel utilization and weighted fairness for data communications in IEEE 802.11 WLAN under the DCF," in Proceedings of Tenth IEEE International Workshop on Quality of Service, pp. 227–236, May 2002.

14. S. Sheu and T. Sheu, "A bandwidth allocation/sharing/extension protocol for multimedia over IEEE 802.11 ad hoc wireless LANs," IEEE Journal on Selected Areas in Communication, Vol. 19, October 2001.

15. I. Aad and C. Castelluccia, "Differentation mechanisms for IEEE 802.11," in Proceedings of IEEE Infocom, 2001.

16. T. Suzuki and S. Tasaka, "Performance evaluation of priority-based multimedia transmission with the PCF in an IEEE 802.11 standard wireless LAN," in Proceedinds of International Symposium on Personal, Indoor and Mobile Radio Communications, Vol. 2, No. 30.

17. L. Jacob, Q. Qiang, R. Radhakrishna Pillai, and B. Pradhakaran, "MAC protocol enhancements and a distributed scheduler for QoS guarantees over the IEEE 802.11 wireless LANs," in Proceedings of Vehicular Technology Conference (VTC), Vol. 4, pp. 2410–2413, September 2002.

18. C. K. Singh, A. Kumar, and P. M. Ameer, "Performance evaluation of an IEEE 802.15.4 sensor network with a star topology," Wireless Network, Vol. 14, pp. 543–568, 2008.

19. F. Shu, "Performance evaluation of the IEEE 802.15.4 CSMA-CA protocol with QoS differentiation," in Proceedings of International Conference on Intelligent Sensors, Sensor Networks and Information Processing, pp. 475–480, December 2008.

20. S. T. Sheu, T. F. Sheu, C. Wu, and J. Y. Luo, "Design and implementation of a reservation-based MAC protocol for voice/data over IEEE 802.11 ad-hoc wireless networks," in Proceedings of IEEE International Conference on Communications (ICC), Vol. 6, pp. 1935–1939, June 2001.

21. S. Sharma, K. Gopalan, and N. Zhu, "Quality of service guarantee on 802.11 networks," in Proceedings Hot Interconnects 9, pp. 99–103, 2001.

22. S. Garg, M. Kappes, and M. Mani, "Wireless access server for quality of service and location based access control in 802.11 networks," in Proceedings of International Symposium on Computers and Communications (ISCC'02), pp. 819–824, 2002.

23. Q. Shen, X. Fang, Pan Li, and X. Fang, "Admission control based on available bandwidth estimation for wireless mesh networks," IEEE Transactions on Vehicular Technology, Vol. 58, No. 5, July 2009.

24. A. Giorgetti, G. Pasolini, and R. Verdone, "Performance evaluation of a channel adaptive wlan polling protocol," in Proceedings of Vehicular Technology Conference, 56th IEEE, Vol. 3, pp. 1379–1383, 2002.

25. M. Gidlund, "An approach for using adaptive error control schemes in wireless LAN with CSMA/CA MAC protocol," in Proceedings of Vehicular Techology Conference (VTC), Vol. 1, pp. 224–228, May 2002.

26. F. Calì, M. Conti, and E. Gregori, "Dynamic tuning of the IEEE 802.11 protocol to achieve a theoretical throughput limit," IEEE/ACM Transactions on Networking, Vol. 8, No. 6, December 2000.

27. A. Koubaa, R. Severino, M. Alves, and E. Tovar, "Improving quality-of-service in wireless sensor networks by mitigating 'hidden-node collisions'," IEEE Transactions on Industrial Informatics, pp. 299–313, Vol. 5, No. 3, August 2009.

28. B. Luca, F. D. Priscoli, T. Inzerilli, P. Mähönen, and L. Muñoz, "Enhancing IP service provision over heterogeneous wireless networks: A path toward 4G," IEEE Communications Magazine, pp. 74–81, August 2001.

29. G. Xylomenos and G. C. Polyzos, "Link layer support for quality of service on wireless internet links," IEEE Personal Communications, pp. 52–60, October1999.

30. S. Jardosh and P. Ranjan, "A survey: Topology control for wireless sensor networks," in Proceedings of IEEE International Conference on Signal processing, Communications, and Networking, pp. 422–427, January 2008.

31. B. Cărbunar, A. Grama, and J. Vitek, "Redundancy and coverage detection in sensor networks," ACT Transactions on Sensor Networks, Vol. 2, pp. 94–128, February 2006.

32. H. Zhai, X. Chen, and Y. Fang, "How well can the IEEE 802.11 wireless LAN support quality of service," IEEE Transactions on Wireless Communications, Vol. 4, pp. 3084–3094, November 2005.

33. H. Levy and M. Sidi, "Polling systems: Applications, modeling, and optimization," IEEE Transactions on Communications, Vol. 38, pp. 1750–1760, October 1990.

34. O. Sharon and E. Altman, "An efficient polling MAC for wireless LANs," IEEE/ACM Transactions on Networking, Vol. 9, pp. 439–451, August 2001.

35. M. Adamou, S. Khanna, I. Lee, I. Shin, and S. Zhou, "Fair real-time traffic scheduling over a wireless LAN," in Proceedings Real-Time Systems Symposium, 22nd IEEE, pp. 279–288, 2001.

36. H. Aydin, R. Melhem, D. Mossé, and P. Mejía-Alvarez, "Optimal reward-based scheduling for periodic real-time tasks," IEEE Transactions on Computers, Vol. 50, February 2001.

37. K. Fall and K. Varadhan, "The ns manual (formerly ns notes and documentation)," [Online]. Available: http://www.isi.edu/nsnam/ns/doc/index.html.

38. S. Wiethölter, C. Hoene, and A. Wolisz, "Perceptual quality of internet telephony over IEEE 802.11e supporting enhanced DCF and contention free bursting," Technical Report TKN-04-11, Telecommunication Networks Group, Technische Universität Berlin, September 2004.

5

CHAPTER

DEVELOPMENT AND EXPERIMENTAL VALIDATION OF DOWNLINK MULTIUSER MIMO-OFDM IN GIGABIT WIRELESS LAN SYSTEMS

Koichi Ishihara*, Yusuke Asai, Riichi Kudo, Takeo Ichikawa, Yasushi Takatori and Masato Mizoguchi

ABSTRACT

Multiuser multiple-input multiple-output (MU-MIMO) has been proposed as a means to improve spectrum efficiency for various future wireless communication systems. This paper reports indoor experimental results obtained for a newly developed and implemented downlink (DL) MU-MIMO orthogonal frequency division multiplexing (OFDM) transceiver for gigabit wireless local area network systems in the microwave band. In the transceiver, the channel state information (CSI) is estimated at each user and fed back to an access point (AP) on a real-time basis. At the AP, the estimated CSI is used to calculate the transmit beamforming weight for DL MU-MIMO transmission. This paper also proposes a recursive inverse matrix computation scheme for computing the transmit weight in real time. Experiments with the developed transceiver demonstrate its feasibility in a number of indoor scenarios. The experimental results clarify that DL MU-MIMO-OFDM transmission can achieve a 972-Mbit/s transmission data rate with simple digital signal processing of single-antenna users in an indoor environment.

Keywords Multiuser MIMO OFDM Downlink Wireless LAN CSI

1. INTRODUCTION

Wireless local area network (LAN) systems have recently seen explosive growth as the prevailing smart devices, e.g., smartphones and tablet computers, deploying wireless LAN chips to enable swift wireless access anywhere. In wireless LANs, broadband services such as video streaming and IP-TV services are becoming pivotal use cases. To provide such broadband services, ever higher data rate is required for wireless LANs. Spectrum efficiency improvement is becoming the most prioritized issue to meet the demand within the limited available frequency channels. Multiple-input multiple-output (MIMO) is one of the key technologies for this purpose; it improves the spectrum efficiency by enabling multiple spatial streams to be simultaneously transmitted from multiple transmit antennas. In the latest wireless access standards, i.e., IEEE 802.11n [1], single-user MIMO (SU-MIMO) with four spatial streams is specified to offer the maximum physical layer data rate of 600 Mbit/s in 40-MHz bandwidth. It also adopts orthogonal frequency division multiplexing (OFDM) to mitigate the effect of multipath fading, resulting in reduced calculation complexity by enabling the MIMO signal processing to be carried out on a per-subcarrier basis. For future wireless communication systems, the multiuser MIMO (MU-MIMO) technique is an attractive candidate to achieve higher spectrum efficiency, especially when the number of antenna branches on the user side is less than that on the access point (AP) [2]. Although SU-MIMO can also improve spectrum efficiency by increasing the number of transmit and receive antennas, the number of antenna branches on the user side is limited in terms of portable device size. In contrast, MU-MIMO makes it possible to multiplex users in the spatial domain so that higher frequency efficiency can be expected even when using simple users. There have been many studies pertaining to MU-MIMO systems, e.g., [3–5]. The next generation wireless LAN system targeting more than 1 Gbit/s is studied in IEEE 802.11ac, and downlink (DL) MU-MIMO-OFDM has been extensively investigated as a means to improve the spectrum efficiency [6, 7]. In computing the transmit beamforming weight at the AP transmitter, higher accuracy is required in DL MU-MIMO transmission than in SU-MIMO transmission. This higher accuracy is required to suppress the inter-stream interference between users since users cannot detect other users' data, especially with the type of simple decoder used in wireless LANs. Furthermore, in IEEE 802.11ac, the transmission bandwidth mandatorily becomes 80 MHz, and the number of transmission antennas is defined as being up to eight, resulting in increased calculation complexity. Therefore, it is important to use implemented hardware to confirm the feasibility of DL MU-MIMO under such actual conditions for wireless LANs. In previous work in this area, an implemented transceiver has been used to experimentally validate SU-MIMO techniques such as eigenbeam space division multiplexing [8–10]. In [11], an implementation scheme of a nonlinear precoding based on Tomlinson-Harashima Procoding for MU-MISO-OFDM has been proposed. This scheme improves the transmission performance by iterative operation.

In this paper, we present indoor experimental results obtained with a newly developed and implemented real-time DL MU-MIMO-OFDM transceiver using field-programmable gate arrays (FPGAs) for the next generation wireless LAN systems. The transmission system consists of an AP with eight antenna elements and six users each having a single antenna element; the transmission bandwidth is 80 MHz. In DL MU-MIMO transmission, channel state information (CSI) is required to compute the transmit beamforming weight at the AP transmitter. In the transceiver, closed-loop feedback is used in accordance with the IEEE 802.11ac Draft. A sounding frame is used to estimate CSI at each user, and each user feeds it back to an AP on a real-time basis. The CSI is used to calculate the transmit beamforming weight at the AP, and the DL MU-MIMO transmission is then carried out via real-time digital signal processing. For computing the transmit beamforming weight in real time, we propose a recursive inverse matrix computation (RIMC) scheme based on a recursive least square (RLS) algorithm [12], in which the inverse of the auto-correlation matrix of the CSI for beamforming weight computation is calculated over successive time intervals. In the proposed scheme, the inverse matrix can be calculated in the arrival order of CSI feedback frames from each user in a time-domain multiple access (TDMA) manner, resulting in reduced overhead for the transmit beamforming weight computation. Moreover, since the transmit beamforming weight obtained by the proposed RIMC scheme is based on zero-forcing (ZF) beamforming, it can ideally provide orthogonal transmission between each stream, and users can decode their own data with a simple decoder in a system such as a single-input single-output (SISO) system. The ZF weight proposed in our paper is a linear precoding scheme that does not require iterative operation, whereas the conventional scheme [11] requires iterative operation to improve the transmission performance. Furthermore, while there is no limitation on the number of transmit antennas to compute the ZF weight in the proposed scheme, the conventional one can compute it only in two-by-two MIMO. Since the maximum number of transmit antennas will be defined as eight in 11ac, the proposed scheme is valid for the next generation wireless LAN. In our previous work [13], we showed experimental results obtained with our developed and implemented transceiver in an experimental room. In this paper, we show not only these results but also two additional experimental results obtained with our developed and implemented transceiver in a large office and a small apartment. The propagation environments are different among these three places. It is important to evaluate the transmission performance of DL MU-MIMO in different environments since the transmit beamforming is affected by the propagation channel. Performance evaluation for time-varying channels is also described as new results. Since the estimated CSI becomes outdated for DL MU-MIMO transmission, the imperfect CSI causes inter-user interference. However, there have been no reports on real-time evaluations of DL MU-MIMO-OFDM transmission performance in time-varying channels. In this paper, we also evaluate the transmission rate obtained with our transceiver in such a channel and verify the feasibility of DL MU-MIMO-OFDM for gigabit-per-second transmission in wireless LAN systems. The paper also mentions information relevant to implementation such as types of FPGA, maximum frequency, and intervals of frame sequence, which show that DL MU-MIMO-OFDM is a promising technology to enable gigabit-per-second transmission for simple terminals with feasible hardware implementation at the AP and users sides.

The remainder of this paper is organized as follows. We explain the transceiver configuration in Section 2. In Section 3, we describe the experimental results we obtained with the transceiver in indoor scenarios. Finally, our conclusions are presented in Section 4.

The following notations will be used throughout this paper: Column vectors (matrices) are denoted by boldface lower (upper) case letters, superscripts 'T' and 'H' respectively stand for transpose and conjugate transpose, and $\mathbf{I}_n$ denotes an $n \times n$ identity matrix.

2. CONFIGURATION OF DL MU-MIMO-OFDM TRANSCEIVER

2.1. Transceiver overview

Figures 1 and 2, respectively, show the DL MU-MIMO-OFDM transmitter and receiver configurations. The major DL MUMIMO-OFDM parameters are listed in Table 1. The carrier frequency is 4.85 GHz, and the maximum number of users for DL MU-MIMO-OFDM transmission is six. In this transceiver, N_S-independent streams are transmitted simultaneously, where N_S is the number of users. The number of subcarriers is 228, which includes 12 pilot subcarriers. At the AP transmitter, the number of antennas N_T is eight, and eight-antenna OFDM signals with 80-MHz transmit bandwidth are generated. At each user's receiver, the signal from the AP is received at an antenna. Channel coding is convolutional coding with $R = 1/2$ coding rate. Digital signal processing at both the AP and user sides is carried out in FPGAs. Table 2 shows a list of the utilized Xilinx Virtex FPGAs (Xilinx, Inc., San Jose, CA, USA) [14] for our implemented DL MU-MIMO. At the AP in our implemented transceiver, six FPGAs are used for signal processing in DL MU-MIMO transmission: a Virtex5 SX95T for weight computation for DL MU-MIMO, three Virtex5 SX95Ts for transmitter signal processing, a Virtex5 FX70T for controlling the previous four Virtex5 SX95Ts and for obtaining CSI, and a Virtex4 SX55 for receiver signal processing. At the users' side in our transceiver, eight FPGAs are used in all: two Virtex5 SX95Ts are used for transmitter signal processing and each processes the transmission signals of the three users, with the processing being carried out independently for each user. Each Virtex4 SX55 is used to process the receiver signals of each user. The maximum clock frequency in our transceiver is 80 MHz.

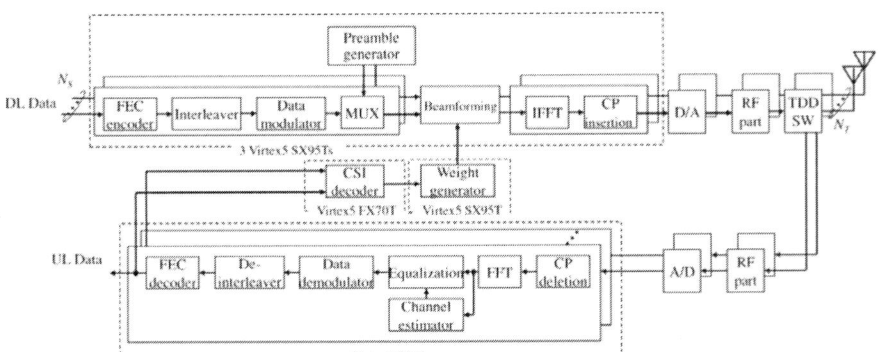

Figure 1. DL MU-MIMO TRANSCEIVER STRUCTURE AT THE AP SIDE.

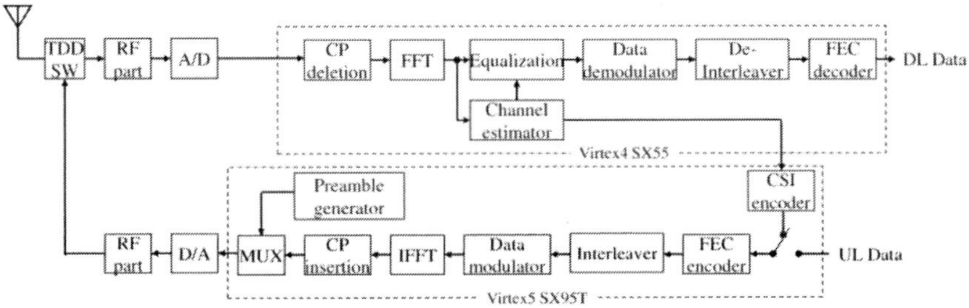

Figure 2. DL MU-MIMO TRANSCEIVER STRUCTURE AT THE USER SIDE.

Table 1. MAJOR PARAMETERS IN DL MU-MIMO-OFDM TRANSCEIVER

Parameter	Unit
Bandwidth	80 MHz
Carrier frequency	4.85 GHz
Number of antennas	8 (AP)/1 (user)
Maximum number of users	6
Maximum total transmit power	30 dBm
Number of subcarriers	228
	(with 12 pilot subcarriers)
OFDM symbol duration	4.0 μs
	(3.2 μs + CP 0.8 μs)
Data modulation	QPSK, 16QAM, 64QAM
Channel coding/decoding	Convolutional coding (R = 1/2)/Viterbi decoding

CP Cyclic prefix.

Table 2. THE LIST OF THE UTILIZED XILINX FPGAS FOR OUR IMPLEMENTED DL MU-MIMO TRANSCEIVER

	Model	Unit	Application	Occupied slices (%)
AP	Virtex5 SX95T	1	Weight computation	90
	Virtex5 SX95T	3	Transmitter signal processing	95
	Virtex5 FX70T	1	Controlling four Virtex5 SX95T, CSI acquisition	67
	Virtex4 SX55	1	Receiver signal processing	83
Six users	Virtex5 SX95T	2	Transmitter signal processing	56
	Virtex4 SX55	6	Receiver signal processing	90

2.2. Frame sequence structure

Figure 3 shows the frame sequence structure in the transceiver. The frame sequence consists of a sounding frame, CSI feedback (CSI-FB) frames, and a data frame. For DL MU-MIMO-OFDM transmission, the CSI from the AP to each user is required to perform the transmit beamforming at the AP's transmitter. In the transceiver, the CSI is estimated using the sounding frame at each user, and it is fed back to the AP.

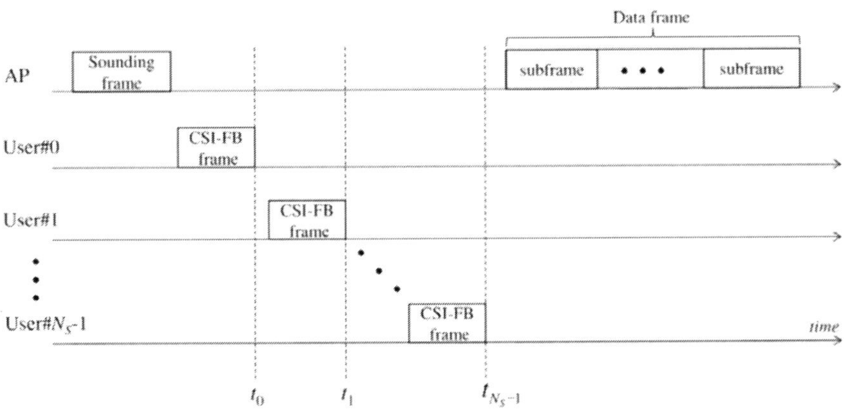

Figure 3. FRAME SEQUENCE STRUCTURE IN THE TRANSCEIVER. The frame sequence consists of a sounding frame, CSI feedback frames, and a data frame.

At the AP, a sounding frame is first transmitted to each user. Figure 4 shows the sounding frame structure. The frame consists of a short preamble and a long preamble. Both preambles are constructed from the 40-MHz design defined in IEEE 802.11n [15] by duplicating and frequency shifting. The long preamble uses an 8 × 8 Hadamard code matrix [16], and eight OFDM symbols are transmitted to estimate the 8 × 1 CSI vector from eight transmit antennas at the AP to an antenna at each user.

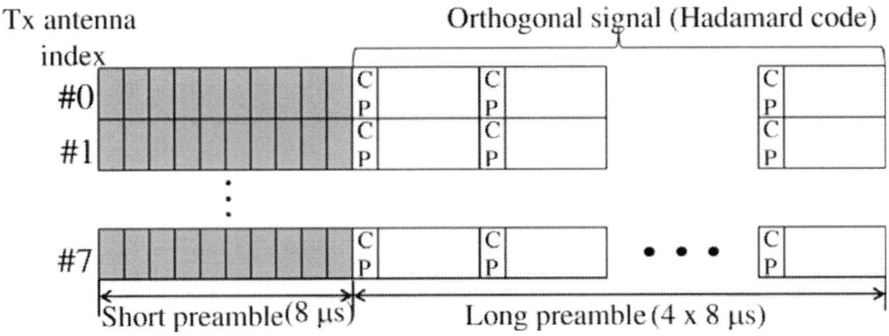

Figure 4. SOUNDING FRAME STRUCTURE. The frame consists of a short preamble and a long preamble.

At each user, the CSI vectors are estimated using the received sounding frame at each subcarrier. In-phase and quadrature-phase quantization are then performed on each CSI estimate. The quantized CSI is encoded using convolutional coding. After interleaving and data modulation mapping, 256-point inverse fast Fourier transform (IFFT) is used to convert the data symbol sequence into an OFDM signal with 228 subcarriers. Then, the OFDM-modulated CSI is transmitted with short and long preambles to the AP as a CSI-FB frame. Here, in uplink transmission, a CSI-FB frame is transmitted from each user in a TDMA manner.

After receiving the CSI-FB frame, the transmit beamforming weight based on ZF beamforming is computed at each subcarrier at the AP for DL MU-MIMO-OFDM transmission. The weight computation scheme will be described in the next section. After computing the beamforming weight, the transmitted DL MU-MIMO-OFDM signal is generated. First, a data sequence is encoded using convolutional coding with the coding rate R = 1/2 and interleaving is carried out for each user. After data modulation and pilot (short and long preambles) insertion, the transmit beamforming is performed in both the data and pilot sequence using the computed weight at each subcarrier. Figure 5 shows the data subframe structure. As the figure shows, the subframe's long preamble consists of $2N_S$ OFDM symbols, and the transmit beamforming for each user is performed in each OFDM symbol to obtain an interference-free CSI estimate at each user. The 256-point IFFT is then applied to produce the time-domain OFDM symbol for each transmitted stream. After cyclic prefix insertion, the DL MU-MIMO-OFDM signal sequences are transmitted from eight antennas at the AP.

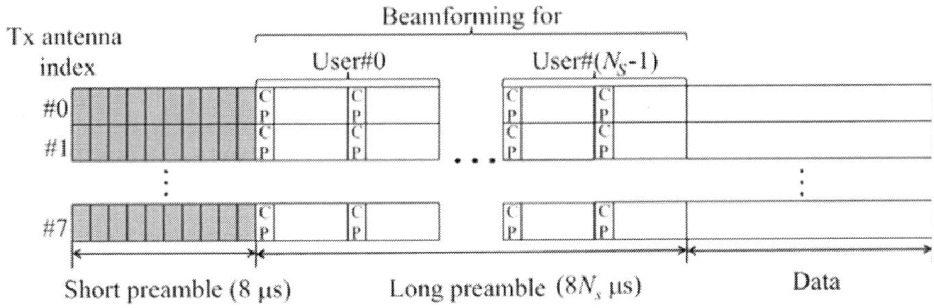

Figure 5. DATA SUBFRAME STRUCTURE. The subframe's long preamble consists of $2N_s$ OFDM symbols, and the transmit beamforming for each user is performed in each OFDM symbol to obtain an interference-free CSI estimate at each user.

At the user, after timing synchronization and frequency offset compensation, channel estimation is carried out using the long preamble at each subcarrier. The channel estimates are used to equalize and demodulate all the data symbols within the subframe. The conventional phase tracking scheme proposed in [17] is also applied to all the data symbols by using the pilot subcarriers. Here, in the transceiver, simple decoding can be carried out as in SISO transmission at each user. This is because ZF-based transmit beamforming can provide orthogonal transmission between each stream ideally, and each user can simply decode the received signal even with a single antenna. After data demodulation, the transmission binary data is obtained through Viterbi decoding, where the number of bits for soft decision at the Viterbi decoder is three due to hardware implementation limitations.

Table 3 shows the duration of each frame and interval. In terms of signal processing, the differences between DL MU-MIMO transmission in 11ac and the conventional transmission in wireless LANs such as that defined in 11n are CSI feedback and DL MU-MIMO signal generation including weight computation. Therefore, we discuss each of these separately. First, we discuss the CSI feedback. Both CSI estimation and CSI-FB frame generation are carried out during the interval (b) in Figure 3, which takes 70 μs. In our transceiver, the CSI-FB frames are transmitted after all the CSI-FB frames for six users are generated for uplink TDMA transmission. Therefore, the actual interval for each user is 70/6 = 11.7 μs. Here, the interval should be less than a short inter-frame space (SIFS) of 16 μs, which is the duration between two successive frames within a frame sequence. Thus, the CSI-FB frames can be transmitted within the SIFS duration. Second, we discuss the DL MU-MIMO signal generation including the weight computation. This is started at the end of the first CSI-FB frame t_0 and finished by the beginning of the data frame. The total signal processing delay for six users is 368 μs. In 11ac, there is no stipulation about this since the CSI-FB frames and DL MU-MIMO data frames can be transmitted independently. However, as the delay becomes longer, transmission performance is degraded since the orthogonality between users is deteriorated due to time-varying channels. We will discuss the effect of interval (d) in the next section.

Table 3. FRAME LENGTHS AND INTERVALS IN OUR IMPLEMENTED DL MU-MIMO TRANSCEIVER

Position	Time (μs)
(a)	44
(b)	70
(c)	60
(d)	318
(e)	$8 + 8N_s + 4n$ ($n = 1,2,\ldots$)
(f)	$440 + 68N_s + 4n$ ($n = 1,2,\ldots$)

2.3. RIMC scheme for DL MU-MIMO-OFDM

In the transceiver, the beamforming weight is used for transmit beamforming. Generally, the weight is given by [18]

$$\mathbf{W}(k) = \mathbf{H}(k)^H \left(\mathbf{H}(k)\mathbf{H}(k)^H + \delta \mathbf{I}_{N_S}\right)^{-1}, \tag{1}$$

where

$$\mathbf{H}(k) = \left[\mathbf{h}_0(k)^T, \ldots, \mathbf{h}_{N_S-1}(k)^T\right]^T. \tag{2}$$

Here, δ is a positive constant used to suppress the noise enhancement. The $\mathbf{H}_m(k)$ term is the $1 \times N_T$ CSI vector between the antenna at the m-th user and the multiple transmit antennas at the AP at the k-th subcarrier and is given by

$$\mathbf{h}_m(k) = \left[H_{m,0}(k), H_{m,1}(k), \ldots, H_{m,N_T-1}(k)\right]. \tag{3}$$

Here, $H_{m,n}(k)$ is the CSI estimate between the n-th antenna at the AP and the antenna at the m-th user at the k-th subcarrier. As shown in Equation (1), to obtain the beamforming weight, it is necessary to compute the inverse of the auto-correlation matrix of the CSI. However, this increases the overhead for the weight computation. In this paper, an RIMC scheme is proposed for the DL MU-MIMO-OFDM transceiver to reduce overhead in computing the beamforming weight [12]. If $\left[\mathbf{H}(k)\mathbf{H}(k)^H + \delta\mathbf{I}_{N_S}\right]$ is nonsingular, Equation (1) can be rewritten as

$$\mathbf{W}(k) = \left(\mathbf{H}(k)^H\mathbf{H}(k) + \delta\mathbf{I}_{N_T}\right)^{-1}\mathbf{H}(k)^H. \tag{4}$$

Here, $\mathbf{R}(k)$ and $\mathbf{R}_m(k)$ are defined as

$$\begin{aligned}\mathbf{R}(k) &= \left(\mathbf{H}(k)^H\mathbf{H}(k) + \delta\mathbf{I}_{N_T}\right)^T \\ &= \sum_{l=0}^{N_S-1}\mathbf{h}_l(k)^H\mathbf{h}_l(k) + \delta\mathbf{I}_{N_T}\end{aligned} \tag{5}$$

$$\mathbf{R}_m(k) = \sum_{l=0}^{m}\mathbf{h}_l(k)^H\mathbf{h}_l(k) + \delta\mathbf{I}_{N_T}, \tag{6}$$

where $\mathbf{R}_{-1}(k) = \delta\mathbf{I}_{N_T}$ and $\mathbf{R}_{N_S-1}(k) = \mathbf{R}(k)$.. The $\mathbf{R}_m(k)$ term can be expressed as the following recursive form:

$$\mathbf{R}_m(k) = \mathbf{R}_{m-1}(k) + \mathbf{h}_m(k)^H\mathbf{h}_m(k) \tag{7}$$

From a formulation of the matrix inversion in [19], we can obtain the inverse matrix of $\mathbf{R}_m(k)$ as

$$\mathbf{R}_m(k)^{-1} = \mathbf{R}_{m-1}(k)^{-1} - \frac{\mathbf{R}_{m-1}(k)^{-1}\mathbf{h}_m(k)^H\mathbf{h}_m(k)\mathbf{R}_{m-1}(k)^{-1}}{1+\mathbf{h}_m(k)\mathbf{R}_{m-1}(k)^{-1}\mathbf{h}_m(k)^H}. \tag{8}$$

Here, the vector $\mathbf{p}_m(k)$ and the scalar $q_m(k)$ are defined as

$$\mathbf{p}_m(k) = \mathbf{R}_{m-1}(k)^{-1}\mathbf{h}_m(k)^H \tag{9}$$

$$q_m(k) = 1 + \mathbf{h}_m(k)\mathbf{p}_m(k) \tag{10}$$

From Equations (9) and (10), Equation (8) is rewritten as

$$\mathbf{R}_m(k)^{-1} = \mathbf{R}_{m-1}(k)^{-1} - \frac{\mathbf{p}_m(k)\mathbf{p}_m(k)^H}{q_m(k)}. \tag{11}$$

Since the CSI-FB frames transmitted from the users are received at the AP in a TDMA manner, calculating the inverse of the auto-correlation matrix progresses in the CSI arrival order. If the time that the CSI of the m-th user obtained is t_m, the calculations of the vector $\mathbf{P}_m(k)$ and the scalar $q_m(k)$ in Equations (9) and (10) are started at t_m, and $\mathbf{R}_m(k)^{-1}$ is updated. According to the above calculation, $\mathbf{R}_m(k)^{-1}$ is carried out iteratively until $m = N_s - 1$. Finally, using the obtained CSI matrix and the inverse of the auto-correlation matrix $\mathbf{R}(k)^{-1} = \mathbf{R}_{N_s-1}(k)^{-1}$ the transmit beamforming weight $\mathbf{W}(k)$ is obtained as

$$\mathbf{W}(k) = \left(\mathbf{R}(k)^{-1}\right)^T \begin{bmatrix} \mathbf{h}_0(k) \\ \vdots \\ \mathbf{h}_{N_S-1}(k) \end{bmatrix}^H. \tag{12}$$

In the proposed RIMC scheme, an inverse matrix computation is carried out recursively by using an RLS algorithm, and the computation circuits of $\{\mathbf{R}_m(k)^{-1}, m = 0,\ldots, N_s - 1\}$ can be communalized. As a result, the circuit size can be reduced. Furthermore, since each CSI-FB frame is transmitted from each user in order in 11ac, the proposed scheme can successively calculate the inverse matrix according to the CSI arrival order from each user. Thus, the AP can start the ZF weight computation at time t_1 in Figure 3 and successively calculate the weight by using every CSI-FB frame. This algorithm can reduce the overhead in calculating the transmit beamforming weight and can hasten the start of the DL MU-MIMO transmission. This leads to small performance degradation due to channel time varying.

3. EXPERIMENTAL RESULTS

Figure 6 shows the indoor measurement environment in an experimental room. The AP antenna arrays had the element spacing of 0.06 m ($\approx 1.0\lambda$, where λ is the wavelength), and the measured value of the spatial correlation between the adjacent transmit antennas was 0.16. The antenna heights for the AP and for the users were, respectively, 1.5 and 1.0 m. The total maximum transmit power was 30 dBm and was allocated to each stream equally. We used linear arrays with eight antennas at the AP and a single antenna at each user. The same assortment of modulation schemes, i.e., QPSK, 16QAM, and 64QAM, was applied to all streams (or users). The coding rate was set at $R = 1/2$. The data frame length is 5.12 ms, and the frames consist of 32 subframes. In the following experiment, the number of feedback information bits for CSI Re[$H_{m,n}(k)$] or Im[$H_{m,n}(k)$] is 12; the CSI-FB uplink frames were transmitted through wired transmission in a TDMA manner, and there is no propagation error in the CSI feedback. Figure 7 shows an example of estimated channel impulse response in the experimental room. We evaluated the channel impulse response by applying inverse discrete Fourier transform (IDFT) to the estimated CSI at each subcarrier in 20-MHz bandwidth since the preamble is constructed from the 40-MHz design defined in 11n to 80 MHz, and there are null subcarriers between subcarrier groups of 20 MHz. Note that the right half of the IDFT output was omitted because it does not show the real channel impulse response. As shown in Figure 7, the measured value of the delay spread in the experimental room is 104.0 ns. Figure 8 plots the average transmission rate as a function of the average total transmit power of DL MU-MIMO-OFDM transmission with the three modulation schemes. The distance between adjacent users is 2.0 m. Here, the transmission rate of DL MU-MIMO-OFDM, r, is defined as

$$r = \sum_{n=0}^{N_s-1} \frac{N_{sc} R \log_2 M}{T_s} (1 - P_n), \tag{13}$$

where N_{sc} is the number of data subcarriers, M is the number of constellation points, T_s is OFDM symbol duration, and P_n is packet error rate (PER) at the n-th user. As the figure shows, the transmission rates of 324, 648, and 972 Mbit/s are achieved at the average transmit power of approximately 3, 10, and 16 dBm with the QPSK, 16QAM, and 64QAM modulation schemes, respectively. These results show that DL MU-MIMO-OFDM transmission with high-level modulation can be applied with a simple decoder at each user in an indoor environment and that a transmission rate over 972 Mbit/s can be achieved.

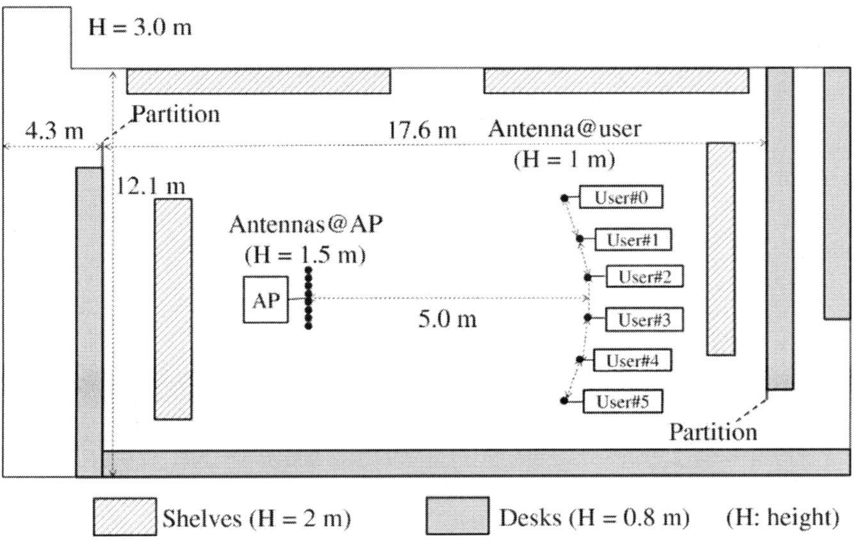

Figure 6. INDOOR MEASUREMENT ENVIRONMENT IN AN EXPERIMENTAL ROOM.

Next, Figure 9 shows the indoor measurement environment in a large office. There were many desks and partitions around the AP and the users, and each user was surrounded by partitions. The antenna heights for the AP and for the users were, respectively, 1.8 and 0.8 m, and the distance between adjacent users was 2.5 m. The AP antenna arrays had the element spacing of 0.03 m (≈0.5λ), and the measured value of the spatial correlation between the adjacent transmit antennas was 0.47. Each user antenna was set up on a desk. The total maximum transmit power was 24 dBm, and the modulation format and coding rate were set to 64QAM and $R = 1/2$, respectively. In this experiment, we measured the transmission performance in two cases. In

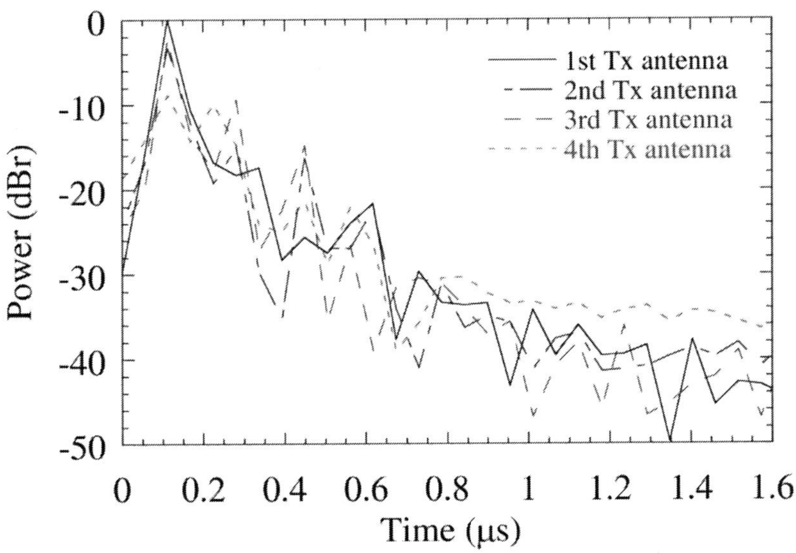

Figure 7 EXAMPLE OF ESTIMATED INSTANTANEOUS IMPULSE RESPONSE BETWEEN AP AND USER IN AN EXPERIMENTAL ROOM. Average delay spread is 104.0 ns.

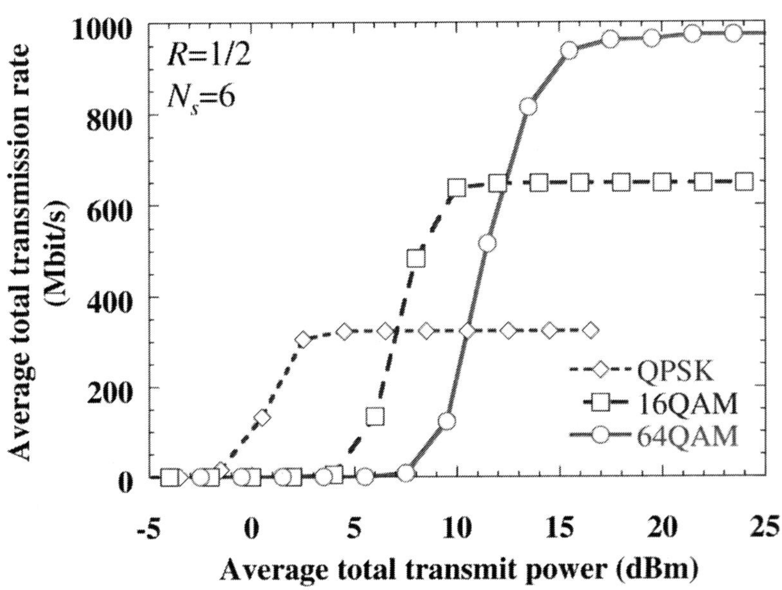

Figure 8. AVERAGE TRANSMISSION RATE. The rate is a function of the total transmit power of DL MU-MIMO transmission with three data modulation schemes.

case 1, people were working with a PC at a desk. In case 2, a person was walking in front of the AP under the case 1 condition. Figure 10 shows an example of estimated instantaneous impulse response in the office. As the figure shows, the measured value of the delay spread in the office is 99.8 ns, which is almost the same as the 104.0 ns for the experimental room. The difference between them is the time-varying characteristic of the channel. The channel in the experimental room is quasi-static, whereas the channel in the office environment is time-varying. Figure 11 plots the average transmission rate as a function of the data frame length. As the data frame becomes longer, transmission performance is degraded since the orthogonality between users is deteriorated due to channel time varying. In particular, the transmission rate in case 2 is lower than that in case 1 because the propagation channel in case 2 varies faster than that in case 1. However, no performance degradation can be seen until after 11.8 and 2.9 ms of the data frame length. These amount to 13.2 and 4.3 ms of the interval (f) in Figure 3 in cases 1 and 2, respectively. This means DL MU-MIMO-OFDM transmission is effective as long as CSI is fed back within these time lengths. Furthermore, the interval (d) of 318 μs is much less than the aforementioned data frame lengths, and the overhead is small. This means the interval is reasonable for DL MU-MIMO transmission.

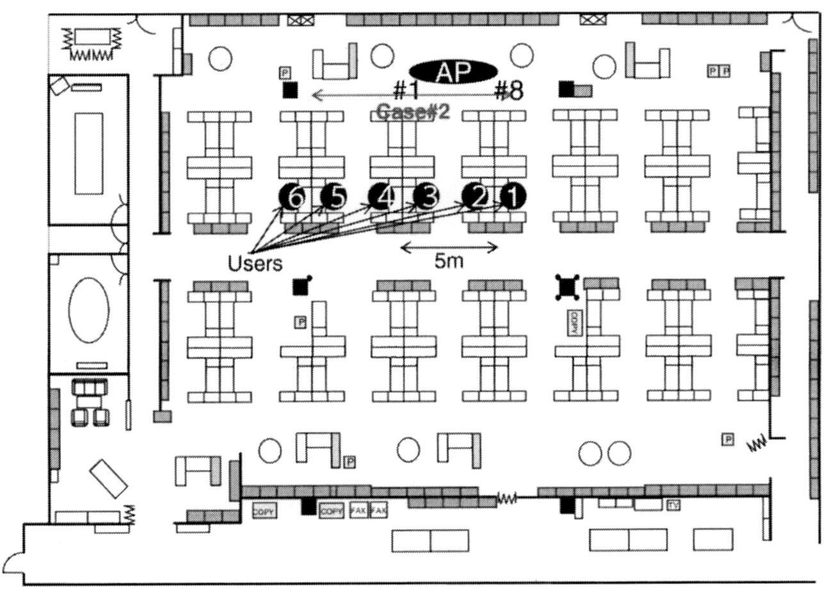

Figure 9. INDOOR MEASUREMENT ENVIRONMENT IN A LARGE OFFICE.

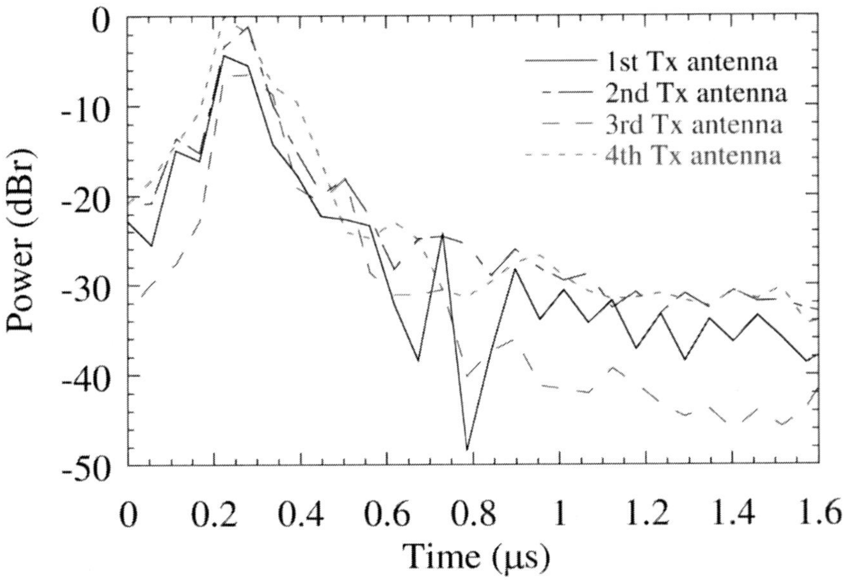

Figure 10. EXAMPLE OF ESTIMATED INSTANTANEOUS IMPULSE RESPONSE BETWEEN AP AND USER IN A LARGE OFFICE. Average delay spread is 99.8 ns.

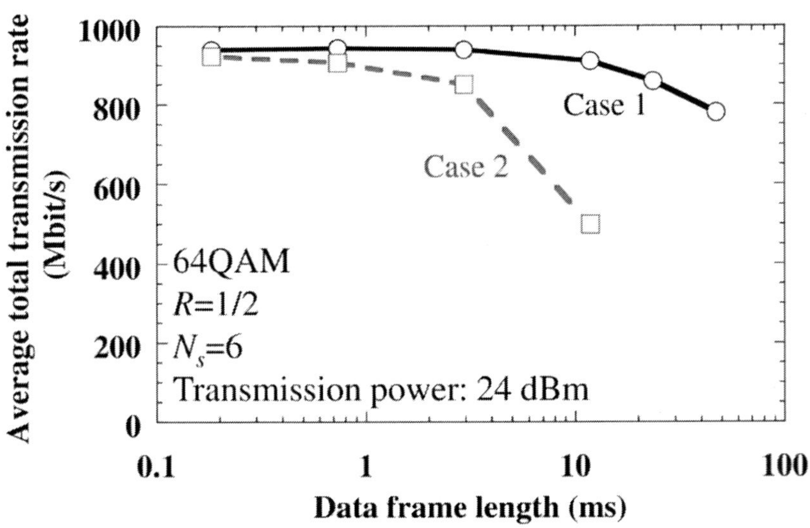

Figure 11. EFFECT OF CHANNEL TIME VARYING. The average transmission rate is a function of the data frame length.

Finally, we measured the transmission rate with the transceiver in a small apartment environment. Figure 12 shows the indoor measurement environment and the user locations in it. The antenna heights for the AP and for the users were 1.5 and 1 m, respectively. The AP antenna arrays had the element spacing of 0.03 m ($\approx 0.5\lambda$), and the measured value of the spatial correlation between the adjacent transmit antennas was 0.54. The total maximum transmit power was 24 dBm, and the modulation format and coding rate were set to 64QAM and $R = 1/2$, respectively. The data frame length is 5.12 ms, and frames consist of 32 subframes. As shown in the figure, the users were located so as to maximize the angle between the AP and the users (i), to minimize it (ii), or to square the user distribution (iii). Figure 13 shows an example of the estimated instantaneous impulse response in an apartment environment. As the figure shows, the measured value of the delay spread in the apartment is 51.6 ns, smaller than that in the office and experimental room environments. This is because the apartment is of wooden construction, and the direct wave through a wall is dominant. Figure 14 plots the cumulative distribution function (CDF) of the transmission rate with the user location and distance as a parameter when there are four users. The modulation scheme and coding rate were set to 64QAM and $R = 1/2$, respectively. When the distance between the adjacent users (d) was 10 cm, the transmission rate in the user location scenario (i) reaches the theoretical maximum transmission rate (648 Mbit/s) at a 10% CDF for this transmission parameter. In contrast, the transmission rates in scenarios (ii) and (iii) are slightly lower. This is because the channel correlation between the users becomes larger than that in scenario (i), and the orthogonality of the transmitted signals between users is deteriorated. However, 520 and 560 Mbit/s transmission rates can be respectively obtained at a 10% CDF in scenarios (ii) and (iii), and they are much larger than those of the TDMA downlink transmission without MU-MIMO. Furthermore, when d reaches 64 cm, the transmission rate in both scenarios (i) and (iii) reaches the theoretical maximum transmission rate, although scenario (iii) is a close-packed structure. Figure 15 plots the CDF of the transmission rate when there are four or six users in scenario (i) when d was 10 cm. As the figure shows, throughout the whole range, the transmission rate is higher for six users than for four, and 948 Mbit/s transmission rates can be achieved with six users at a 50% CDF. Therefore, the DL MU-MIMO is also valid in such an environment with a small delay spread. The total transmission rate with four users reaches the theoretical maximum transmission rate at a 10% CDF. In contrast, the total transmission rate with six users reaches it at a 95% CDF. However, the total (per user) transmission rates of 640 (160) and 780 (130) Mbit/s with four and six users can be achieved at a 10% CDF, and it is found that the number of multiplexed users increases and the total transmission rate improves.

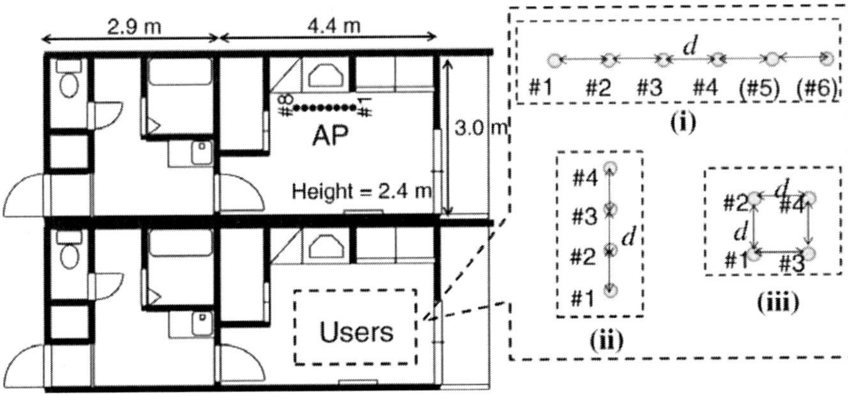

Figure 12. INDOOR MEASUREMENT ENVIRONMENT IN A SMALL APARTMENT.

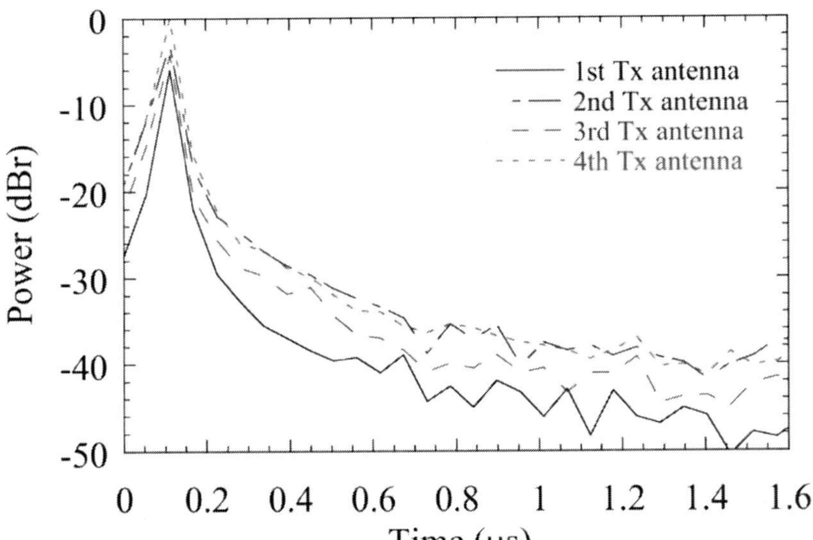

Figure 13. EXAMPLE OF ESTIMATED INSTANTANEOUS IMPULSE RESPONSE BETWEEN AP AND USER IN A SMALL APARTMENT. Average delay spread is 51.6 ns.

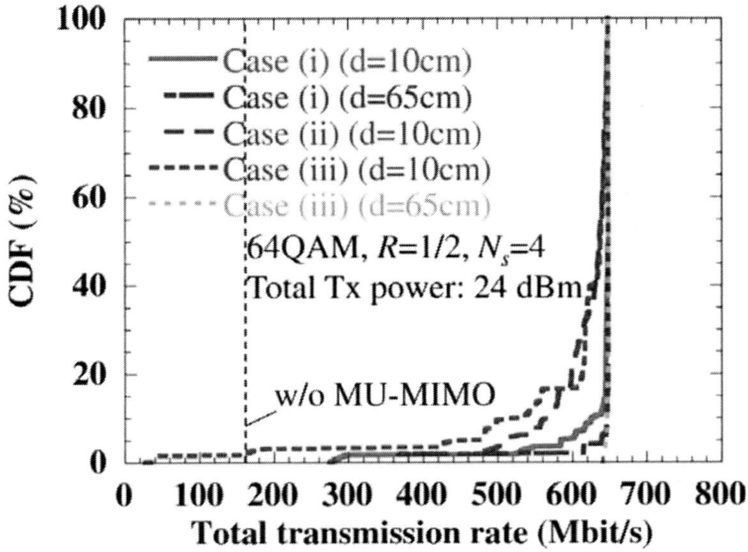

Figure 14. EFFECT OF USER LOCATIONS. CDF of the transmission rate with the user location and distance as parameters when there are four users.

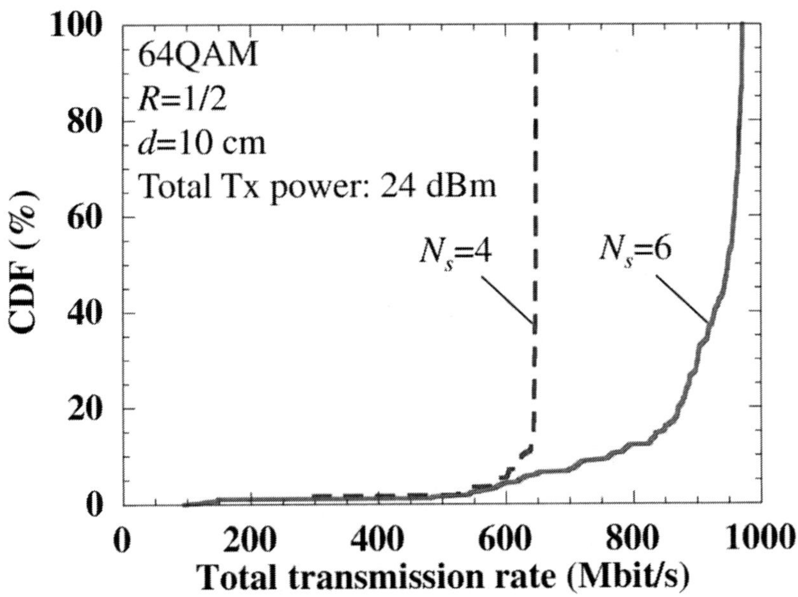

Figure 15. EFFECT OF THE NUMBER OF USERS. CDF of the transmission rate when there are four or six users in scenario (i) when *d* was 10 cm.

4. CONCLUSION

This paper presents the indoor experimental results obtained for real-time DL MU-MIMO-OFDM transmission using a RIMC scheme in gigabit wireless LAN systems. In the proposed scheme, the inverse of the auto-correlation matrix can be calculated in the arrival order of CSI feedback frames from each user in a TDMA manner, resulting in reduced overhead for the transmit beamforming weight computation. In this paper, we evaluated the effects of the delay spread of the propagation channels and the time variations in three places with different channel characteristics and verified the feasibility of DL MU-MIMO-OFDM for gigabit-per-second transmission in wireless LAN systems. The experiment results showed that real-time DL MU-MIMO transmission of 972 Mbit/s was achieved with zero forcing-based transmit beamforming obtained by the proposed RIMC scheme at the AP and a simple decoder with each single-antenna user. In measuring the transmission rate in actual indoor environments, we found that a total transmission rate of 780 Mbit/s was achieved with six users at a 10% CDF. The obtained results confirm that DL MU-MIMO-OFDM is a promising technology to enable gigabit-per-second transmission for simple terminals with feasible hardware implementation at the AP.

AUTHORS' AFFILIATIONS

(1) NTT Network Innovation Laboratories, NTT Corporation

REFERENCES

1. *Perahia E: IEEE 802.11n development: history, process, and technology. IEEE Commun. Mag. 2008, 46(7):48-55.*
2. *Goldsmith A, Jafar SA, Jindal N, Vishwanath S: Capacity limits of MIMO channels. IEEE JSAC 2003, 21(5):684-702.*
3. *Spencer QH, Swindlehurst AL, Haardt M: Zero-forcing methods for downlink spatial multiplexing in multiuser MIMO channels. IEEE Trans. Signal Processing 2004, 52: 461-471. 10.1109/TSP.2003.821107*
4. *Choi LU, Murch RD: A transmit preprocessing technique for multiuser MIMO systems using a decomposition approach. IEEE Trans. Wireless Commun. 2004, 3: 653-668.*

5. *Shen Z, Chen R, Andrews JG, Heath RW Jr, Evans BL: Low complexity user selection algorithms for multiuser MIMO systems with block diagonalization. IEEE Trans. Signal Processing 2006, 54: 9.*
6. *802.11 Working Group of the IEEE 802 Committee: P802.11ac/D5.0, Draft Standard for Information Technology—Telecommunications and Information Exchange Between Systems: Local and Metropolitan Area Networks—Specific Requirements Part 11: Wireless LAN Medium Access Control (MAC) and Physical Layer (PHY) Specifications: Enhancements for Very High Throughput for Operation in Bands below 6 GHz. New York: IEEE; 2013.*
7. *Nee RV: Breaking the gigabit-per-second barrier with 802.11ac. IEEE Wireless Commun. 2011, 18(2):4.*
8. *Katser T, Wtlzeck A, Berentsen M, Rupp M: Prototyping for MIMO systems—an overview. Vienna, Austria: Paper presented at the 12th EUSIPCO; 2004.*
9. *Font-Bach O, Bartzoudis N, Pascual-Iserte A, Bueno DL: A real-time MIMO-OFDM mobile WiMAX receiver: architecture design and FPGA implementation. Comp. Netw. 2011, 55(16):3634-3647. 10.1016/j.comnet.2011.02.018*
10. *Caban C, Mehlfuhrer C, Langwieser R, Scholtz AL, Rupp M: Vienna MIMO testbed. EURASIP J. Appl. Signal Proces. 2006, 2006: 1-13.*
11. *Ludwig F, Budweg A, Paul S: FPGA implementation of ZF-THP for MU-MISO-OFDM systems. Essen, Germany: Paper presented at the 17th international OFDM workshop 2012 (InOWo '12); 2012.*
12. *Monzingo RA, Miller TWM: Introduction to Adaptive Arrays. New York: Wiley; 1980.*
13. *Ishihara K, Asai Y, Kudo R, Ichikawa T, Mizoguchi M: Indoor experiments on real-time multiuser MIMO transmission in wireless LAN systems. Paris, France: Paper presented at IEEE Wireless Communications and Networking (WCNC); 2012.*
14. *Xilinx Inc: The Xilinx Inc. homepage. 2013.. Accessed 18 March 2013*
15. *IEEE Computer Society: IEEE Standard for Information Technology—Telecommunications and Information Exchange Between Systems: Local and Metropolitan Area Networks—Specific Requirements Part 11: Wireless LAN Medium Access Control (MAC) and Physical Layer (PHY) Specifications: Enhancements for Higher Throughput. New York: IEEE; 2009.*

16. *Proakis JG: Digital Communications. 4th edition. New York: McGraw-Hill; 2001.*
17. *Robenson P, Kaiser S: Analysis of the effects of phase-noise in orthogonal frequency division multiplex (OFDM) systems. Seattle, USA: Paper presented at the IEEE international conference on communications (ICC); 1995.*
18. *Joham M, Utschick W, Nossek JA: Linear transmit processing in MIMO communications systems. IEEE Trans. Signal Proces. 2005, 53(8):2700-2712.*
19. *Lutkepohl H: Handbook of Matrices. New York: Wiley; 1996.*

6

CHAPTER

A NEW SCHEME TO IMPROVE OVERALL TCP THROUGHPUT WITH VERTICAL HANDOVER BETWEEN 3G CELLULAR PACKET NETWORKS AND WIRELESS LANS

Chang Hyeon Lim1 and JuWook Jang[2]

ABSTRACT

One major difference between 3G cellular networks and Wireless LANs is in how packet losses are dealt with. 3G cellular networks employ selective repeat ARQ (SR-ARQ) which may result in packet-reordering due to local retransmissions. Wireless LANs employ stop-and-wait ARQ which may drop some packets in case the buffer overflows. While much work has been performed for TCP throughput enhancement with vertical handover between wired networks and wireless networks, little attention is paid to the vertical handover between two different wireless networks. It is expected for majority of new mobile devices to be able to connect to both 3G cellular networks and wireless LANs by being equipped with multiple wireless interfaces. If this is the case, on-going TCP transmission may move between 3G cellular networks and wireless LANs. In this paper, we propose a scheme to improve TCP throughput with vertical handoff between 3G cellular networks and wireless LANs. For this we develop an algorithm which differentiates packet-reordering from packet losses and congestion losses from wireless losses. Simulation of our algorithm using ns-2 shows significant increase in throughput as large as 80% compared to TCP-SACK. The proposed scheme is able to respond to the change of loss pattern and then improve the TCP throughput during handover.

1. INTRODUCTION

Vertical handover [1] refers to the process of changing interfaces for a mobile node with multiple network interfaces to move from one network to another network. Recent mobile devices with two or more interfaces are able to connect to different types of networks. Previous researches on vertical handover have addressed the handling of the difference in bandwidth-delay products between two different types of networks (e.g., 3G cellular networks and WLANs) [2–4]. Known TCP enhancement schemes over wireless packet losses [5–8] have paid little attention to vertical handover cases. Differentiation of congestion losses from wireless losses over a network which consists of a wired part and a wireless part is performed in [5, 6], and the detection of packet-reordering due to the retransmission at the link layer on a wireless last hop is dealt with in [8].

In this paper, we consider a new problem that an on-going TCP flow on a mobile node may experience packet losses of different natures over vertical handover depending on link layer retransmission schemes involved. For example, the packet can be lost due to wireless error over WLAN or due to congestion loss over the wired part. Or packets may be regarded as loss due to reordering which happens over 3G cellular networks due to its recovery mechanism. To obtain best TCP throughput over vertical handover, we need to accurately identify these three different causes of packet loss, congestion loss, wireless loss, and packet-reordering.

To address this problem, we propose a new TCP congestion control scheme which adapts to the link layer transmission schemes involved in vertical handover. Specifically, we deal with handover between 3G and WLAN which, we expect, would be most prevailing in the near future mobile environment. The proposed scheme is a modification of loss differentiation (LD) scheme [7] to improve its performance over handover between 3G and WLAN. The proposed scheme deals with wireless/wireless handover problem, while the original LD differentiates wireless loss from congestion loss and thus with steady-state wired/wireless problem. If we employ the LD in its original form for this new problem, the LD does not function as expected. The key parameters of LD have been tuned over time to the network and do not fit new network immediately after the handover. TCP-DCR (Delayed Congestion Response) [8] works well over 3G cellular networks, but not for handover. F-RTO (Forward Retransmission Timeout) [9] can be used over handover. However, the performance of proposed scheme excels the F-RTO. The proposed scheme significantly improves any of the algorithms mentioned above for handover from WLAN to 3G cellular packet network or vice versa. Figures 11, 12, 13, and 14 show the simulation results before handover and after handover for various scenarios. For example, our scheme (proposed scheme II) excels DCR by 80% in throughput averaged for 250 seconds after handover. Furthermore, our scheme outperforms the simple combination of LD and DCR (proposed scheme I) by about 60% in throughput averaged for 250 seconds after handover.

We summarize our contribution as follows. Proposed schemes are not mere combinations of known algorithms (LD, TCP-DCR and/or FRTO). First, we solve a new problem. We try to solve the problem of TCP throughput degradation over vertical handover between 3G cellular packet networks and Wireless LANs, while the original LD [7] is developed for TCP improvement for wired/wireless paths. In other words, our scheme deals with wireless/wireless handover problem, while the original LD deals with steady-state wired/wireless problem.

Second, we do not combine three algorithms in their original forms. For example, if we employ the LD in its original form for this new problem, the LD does not function as expected. On handover, the key parameters of LD (RTTmin, RTTavg, RTTdev, etc.) might not closely reflect the current condition of the new wireless network for some time. The reason is that the parameters have been tuned to the old wireless network before handover and therefore are not supposed to be used for the new wireless network. We modified the LD to deal with the problem of LD over handover. Similarly, TCP-DCR [8] works well over 3G cellular networks, but not for handover. F-RTO [9] can be used over handover. However, our scheme overlaps with F-RTO only in the detection of handover. We may avoid using F-RTO by using one of known cross-layer approaches [9] for detection of handover. We choose to incorporate the handover detection part from the F-RTO since we want to retain layer independency (so that our scheme may work independently from link layer).

Third, our scheme significantly improves any of the algorithms mentioned above for handover from WLAN to 3G cellular packet network or vice versa.

Figures 11 through 14 show the simulation results before handover and after handover for various scenarios.

The rest of this paper is organized as follows. We define the problem of TCP throughput enhancement over vertical handover between 3G networks and WLAN in Section 2. Related works are summarized in Section 3. Proposed schemes are presented in Section 4. The performance evaluation of our schemes is provided in Section 5. Section 6 concludes the paper.

2. PROBLEM DEFINITION

The retransmission technology at a wireless link layer can hide the packet loss from the upper layer, for example, TCP. Despite this retransmission, packet losses are still exposed to the TCP layer [10, 11] as shown in Figure 1. The packet losses can reduce the overall TCP throughput no matter where they come from. We consider two types of network: IEEE 802.11-based wireless LAN and 3G packet network.

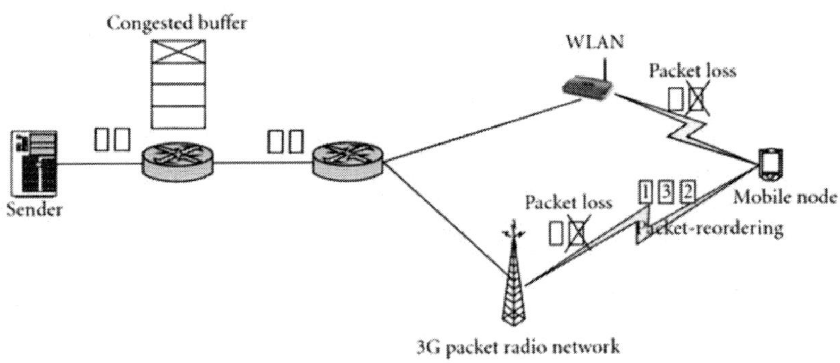

Figure 1. THREE PACKET LOSS SCENARIOS: (1) CONGESTION LOSSES OVER THE WIRED NETWORK, (2) PACKET LOSS OVER WLAN, AND (3) PACKET LOSS AND PACKET REORDERING OVER 3G PACKET NETWORK.

IEEE 802.11-based wireless LAN adopts the Stop-and-Wait ARQ [12] into their link retransmission technique. In this scheme, when a packet is lost, the sender (access point) retransmits the lost packet several times to a mobile node. The packet loss is exposed to the TCP layer when the number of retransmission trials exceeds the constraint [11]. Then, the TCP sender corresponding to the mobile node halves its congestion window. This leads to reduced TCP throughput.

3G packet network can hide most of packet losses to TCP layer due to selective-repeat ARQ [13, 14]. However, this retransmission technology may generate out-of-order packets. On packet-reordering, the TCP sender retransmits the lost packet and reduces its sending window even though the receiver receives all of packets without losses. Some packets are still lost even with packet loss rate of 1% due to three reasons: queue overflow at link layer, timeout, and the number of transmitting packets more than the size of the retransmission queue [10].

A mobile node with two or more different types of the link layer interface (e.g., wireless LAN and 3G packet network) needs to be able to handle the packet loss and packet-reordering problem. Many known schemes on the packet losses over wireless links mainly deal with classification of the causes of packet losses [6, 7]. They have improved TCP throughput over wired and wireless network. Packet-reordering due to the advanced link layer retransmission is considered as a loss by TCP [8]. Generally, end-to-end loss differentiation schemes could keep their TCP congestion window against packet-reordering. However, they may redundantly retransmit the packet due to packet-reordering.

Our assumption on the packet loss based on literatures is as follows:

1. packet loss model over WLAN: Gilbert 2-state Markov chain [15],
2. packet loss model over 3G: uniformly random loss,
3. packet-reordering over 3G:

(a)uniformly random event,

(b)swapping a packet for the next packet [8, 16].

3. RELATED WORK

The delayed congestion response scheme (TCP-DCR) [8] distinguishes packet-reordering from congestion losses over 3G networks. This scheme waits for the normal ACK to handle packet-reordering to reduce unnecessary retransmission on three duplicate ACKs generated due to reordered packets. However this scheme cannot distinguish wireless losses from congestion losses. Actually, packet losses as well as packet-reordering happen on 3G network.

The loss differentiation schemes differentiate congestion losses from wireless losses [6, 7]. These schemes improve TCP throughput over wireless losses by classifying the cause of packet losses. However, they also classify packet-reordering into packet losses, and therefore retransmit unnecessary packets.

IEEE 802.21 framework [17] deals with vertical handover: a mobile node discovers a new access link, registers itself to the new network, disconnects itself from the old access link, and communicates with corresponding host over the new access link. However, IEEE 802.21 framework does not include the transport layer to support vertical handover. TCP is still exposed to losing packets due to difference among access links and several sources of packet losses (e.g., congestion, wireless errors, and packet-reordering).

Recent research on TCP modifications for vertical handover has focused on compensating for the difference of link delay or bandwidth-delay product between two different wireless links [2–5]. These problems arise when a mobile node moves to the network with a link layer technology different from the current network. If a mobile node moves to the network with a smaller bandwidth-delay product, the mobile node may lose packets since its current congestion window is larger than the new bandwidth-delay product.

In [2], a mobile node estimates the bandwidth-delay products (BDP) of the both paths before and after the handover. After performing the vertical handover procedure, the mobile node then returns the ACK at the rate according to the bandwidth of the new wireless access link. When the BDP becomes smaller, the TCP receiver adjusts congestion window (cwnd) by sending the triple duplicate acknowledgements (ACKs). When the BDP becomes larger, the receiver returns many ACKs to increase the cwnd rapidly. This compensation needs supports of link layer or of the access node of wireless network. In [3], network layer controls the route of ACKs during vertical handover. In particular, for handover from a fast link to a slow link, the ACKs of the first few slow-link packets are routed to the old fast link to reduce the gap between old RTT and new RTT. For handover from a slow link to a fast link, similar procedure is performed. In [4], the TCP receiver sets the *awnd* (acknowledged window) to the minimum value of the receiver window size and the estimation of the maximum BDP over GPRS network at the receiver-side, and reports it to the sender. The maximum BDP is calculated as follows: the TCP receiver counts the amount of TCP data received from the instant; a timestamp is put on an acknowledgement (ACK) to the instant; the TCP packet carrying the echoed timestamp has been received [4]. This scheme follows end-to-end semantics and can be easily implemented at mobile node.

There has been a TCP solution to facilitate smooth handover with the initialization of the TCP sender between two different types of networks [5]. Their TCP solution is for a TCP receiver to trigger the vertical handover event, and for the corresponding TCP sender to initialize the congestion window, slow-start threshold, and retransmission timer on the vertical handover event. This scheme can help a TCP flow adjust its congestion window to the changed BDP.

These previous researches have focused on only the difference of bandwidth-delay product (BDP) between two different packet networks when the vertical handover occurs. The shrinking of BDP increases the number of dropped packets in proportion to the difference between the size of TCP's congestion window and the size of the new BDP. We know that the BDP can be changed when a mobile node connects to a new network with a link layer different from the previous network. This problem should be addressed.

However, the focus of our paper is not on the adaptation of TCP for the change in the BDPs of the links but on the adaptation of TCP for the change in the link-level retransmission schemes on handover. Moreover, we are not alone in assuming the same fixed delay in wired part of both paths in simulation setup for TCP adaptation over vertical handoff [18]. We assume that the change of BDP over the vertical handover does not happen, or that, at least, the difference of the BDPs has been compensated. According to [19], the BDP of 3G cellular network and WLAN could be similar as 15 Kbytes and 14 Kbytes, respectively.

4. PROPOSED TCP MODIFICATION

In this section, two new TCP congestion control schemes to solve TCP throughput degradation over vertical handover are proposed. Proposed scheme I (also called DCR-LD) is basically a combination of our loss differentiation (LD) scheme [7] and the TCP-DCR [8]. Proposed scheme II (also called DCR-FRTO-LD) is a modification of LD which borrows handover detection algorithm from the Forward-Retransmission Timeout (F-RTO) [9] and incorporates the TCP-DCR scheme for improved throughput over 3G.

Proposed scheme I improves TCP throughput over wireless packet losses and packet-reordering. On receiving three duplicate ACKs, the DCR part of the proposed scheme I differentiates real packet losses from packet-reordering. On duplicate ACK, it waits for a new ACK for the duration one estimated RTT. If a TCP sender receives a new ACK, the duplicate ACK is considered due to packet-reordering, and the TCP sender keeps its current window size. Otherwise, the loss differentiation scheme is invoked in an estimated RTT to identify the cause of the loss. If the measured RTT just before the duplicate ACK is lower than the threshold, the loss is classified as wireless loss, and then the congestion window is allowed to retain the current value. Otherwise, the TCP sender regards the loss as congestion loss and halves the congestion window. Consequently, proposed scheme I is able to identify the three types of packet losses, packet-reordering, wireless loss, and congestion loss.

F-RTO [9] is for the TCP sender to keep the congestion window with detection of delay spike due to vertical handover and to reset the measured RTT and its deviation to initial values. When a timeout event happens, F-RTO algorithm sends two probe packets of new sequence numbers. If a normal ACK is received, F-RTO assumes that the vertical handover has occurred, and the TCP timeout retransmission procedure is cancelled. Otherwise, the retransmission procedure is invoked. We may avoid using F-RTO in our proposed scheme II by using one of known cross-layer approaches [18] for detection of handover.

In proposed scheme II, the measured RTT, its deviation, and the minimum RTT are reset to new measured values to prepare LD into the new wireless environment. The current congestion window is retained. Furthermore, LD is not invoked for several rounds until the parameters needed for LD are tuned to the new wireless link.

For completeness of presentation, we present our loss differentiation scheme [7] and its proposed modification for handover in Section 4.1. The modified scheme is used in proposed schemes I and II.

4.1. Loss Differentiation Scheme and Its Modification for Handover

In Section 4.1.1, we briefly review our loss differentiation scheme [7] which is originally proposed to infer the cause of each packet loss as congestion loss over wired part or wireless loss for wired/wireless networks. Then the modification of LD for vertical handover is presented in Section 4.1.2.

4.1.1. Loss Differentiation Scheme

Let T_{cur}denote the RTT measured immediately before the current packet loss. It is used in (1) as an indicator. The RTT consists of the propagation delay and the queueing delay [10]. Let $T_{avg}$and T_{dev}denote an exponentially weighted moving average of RTT and the deviation, respectively. The initial values of T_{avg} and T_{dev} are set to 0 and 0.5 s, respectively. They are updated by $T_{avg} = (7/8)T_{avg} + (1/8)T_{cur}$ and $T_{dev} = (3/4)T_{dev} + (1/4)|T_{avg} - T_{cur}|$. Let T_{min} denote the propagation delay which is reset by the expiration of the retransmission timer. The current packet loss is determined to be a congestion loss if (1) is satisfied:

$$\frac{T_{cur} - T_{avg}}{T_{dev}} > 2\left(\frac{T_{min}}{T_{cur}}\right)^k - 1. \tag{1}$$

We assume that the connection between the source and the destination is on a fixed path throughout the connection's lifetime.

The LD [7] accurately classifies congestion and random losses over varying network parameters, that is, buffer size, wireless random loss rate, and so forth. This scheme can reliably classifies the congestion loss with an accuracy close to 100% even with a small buffer of size the 20% of the bandwidth-delay product and has less dependency on network parameters compared with other schemes in literature.

4.1.2. Proposed Modification of Loss Differentiation Scheme for Handover

In the original LD [7], we have assumed that the connection between the source and the destination is on a fixed path throughout the connection's lifetime. However, on vertical handover, a mobile node may change its last wireless hop to a different link layer. This implies that the characteristics of the path, that is, propagation delay and the cause of packet losses, may change. We modify the LD in such a way that, on vertical handover, the parameters for the loss differentiation are reset for LD to accurately infer the cause of packet losses on the changed wireless link.

We show via simulations the efficiency of the modified loss differentiation scheme when the vertical handover occurs. The parameters setup and the environment are the same as Table 1 in Section 5. Since the loss differentiation depends on the measured RTT and its statistical values, it is important how soon statistical values are to become stable after handover. Equation (2) is derived from (1):

Table 1. Simulation Configurations.

Term	3G cellular	WLAN
Packet loss model	Uniform random loss model with rate 0.5%	Gilbert Markov 2-state model with good state 0.1%, bad state 10%, and transition rate 0.5
Bandwidth	1 Mbps	10 Mbps
Delay	200 ms	20 ms
Packet-reordering model	Uniform and random packet-reordering probability of 5%. Swap a packet for the next packet. (5% is chosen out of the channel error rate range from 0% to 8% in [8, Figure 5, page 523])	—
	Wired part	
Congestion loss model		Forward cross traffic (10 connections) with congestion loss rate 1.5%
Bandwidth		20 Mbps
Delay		1 ms
Router buffer		30 packets
	Vertical Handover	
Vertical handover Event		At 250 s, simulation is paused during 30 seconds

$$T_{\mathrm{cur}} > T_{\mathrm{avg}} + T_{\mathrm{dev}}\left(2\left(\frac{T_{\mathrm{min}}}{T_{\mathrm{cur}}}\right)^k - 1\right). \quad (2)$$

The left-hand side of (2) is the measured RTT. The right-hand side of (2) is the threshold of loss differentiation. Each term in (2) is updated on receipt of an acknowledgement packet. A packet loss is regarded as a congestion loss if the measured RTT(T_{cur}) is greater than the threshold, otherwise, a wireless loss.

We modifies the retransmission timeout procedure in TCP to set statistical values, for example, $T_{avg}$and T_{dev}, to the initial values on vertical handover, and to set the propagation delay T_{min} to a large value. We monitor the measured RTT and the threshold. The purpose of this monitoring is to determine if the modified loss differentiation scheme accurately identifies congestion loss even when the statistical values are yet to be stabilized after handover.

In our simulation scenario, it is assumed that the spurious timeout during vertical handover is set to be equal to 3 seconds and that the bandwidth-delay product of 3G network is the same as WLAN (the reasoning behind this assumption is provided by the last paragraph of Section 3). The timeout event of handover is invoked by pausing the simulation for several seconds. To simulate 3G links, bandwidth and delay are set to 1 Mbps and 200 milliseconds, respectively. To simulate WLAN, bandwidth and delay are set to 10 Mbps and 20 milliseconds. There is no cross traffic and no wireless packet loss.

Figure 2 shows the vertical handover from 3G network to WLAN. The loss differentiation scheme correctly indicates noncongestion when T_{cur} is close to T_{min} and congestion when T_{cur} is over the threshold in (2). Actually, this case is to simulate that the mobile moves from a slow link to a fast link. Thus, T_{min} adapts into a lower value over handover to the faster link.

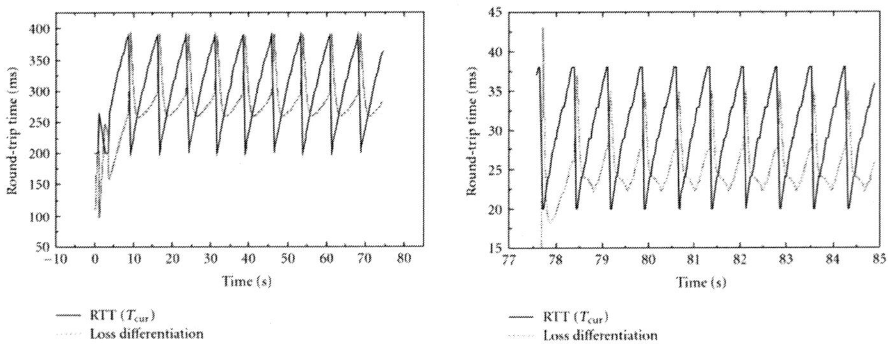

Figure 2. LOSS DIFFERENTIATION ON VERTICAL HANDOVER FROM 3G NETWORK TO WLAN. Before handoverAfter handover.

The case of the vertical handover from WLAN to 3G network needs requires that the minimum RTT T_{min} is set to a great initial value on handover. The original loss differentiation scheme [7] may wrongly classify the cause of packet losses on handover due to a large difference of minimum RTT between a fast link and a slow link. As shown in Figure 3, the modified loss differentiation scheme works well in the case of the vertical handover from WLAN to 3G.

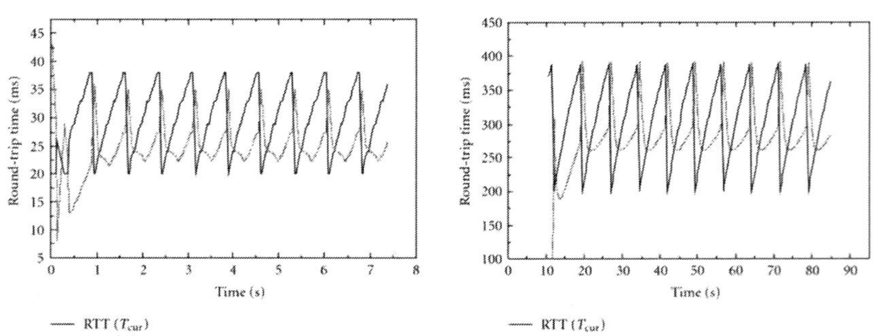

Figure 3. LOSS DIFFERENTIATION ON VERTICAL HANDOVER FROM WLAN TO 3G NETWORK. Before handoverAfter handover.

4.2. Proposed Schemes

The proposed scheme I is a combination of LD [7] and TCP-DCR [8]. The proposed scheme II is a modified loss differentiation scheme (introduced in Section 4.1.2) which borrows handover detection scheme from F-RTO [9] and incorporates TCP-DCR [8] for performance improvement for 3G.

Figure 4 shows the state transition diagram of the proposed Scheme II. The diagram includes states for reset, slow-start, fast recovery, congestion avoidance, F-RTO Transmit Two Packets (handover detector borrowed from F-RTO), DCR (distinguishes between packet-reordering and packet losses via delayed retransmission), and the modified loss differentiation. The congestion window starts from the reset state and then grows in slow-start state.

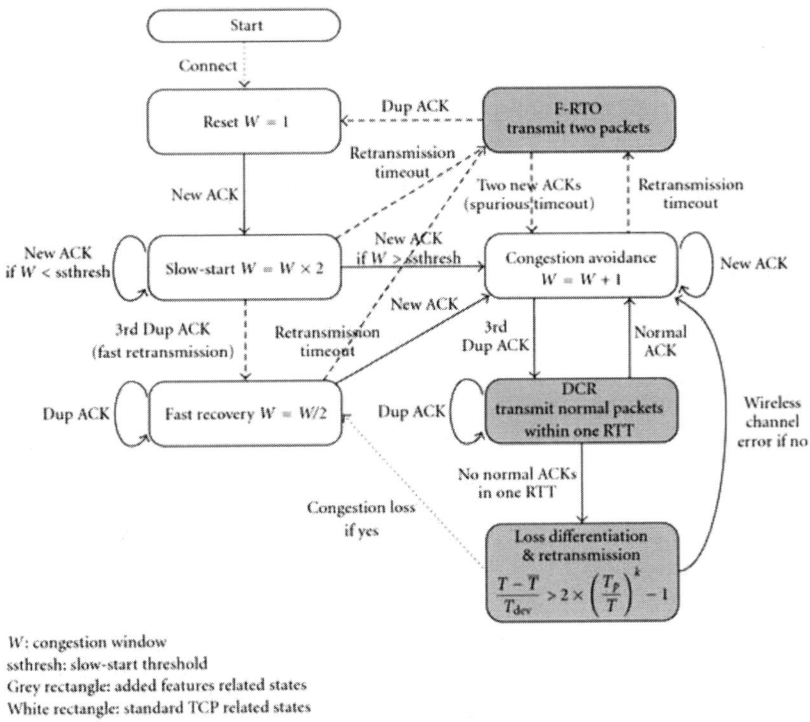

Figure 4. PROPOSED TCP MODIFICATION IN RESPONSE OF VERTICAL HANDOVER. After returning congestion avoidance from F-RTO, loss differentiation should not be involved by 3rd Dup ACK for *n* rounds.

The event of three-duplicate ACKs leads to fast recovery of halving the congestion window, and then the state moves to congestion avoidance state. Timeout leads to F-RTO state which is to avoid delay spike due to vertical handover. F-RTO [9] retransmits the packet that triggered timeout on entering retransmission timeout and waits for the ACK. The sender's receipt of two new ACKs aloows the sender to transmit new packets continuously in congestion avoidance state. However, on receiving one or two duplicate ACKs, the state goes to reset. Whenever the state goes to F-RTO, T_{min} is set to a large value and is updated per every RTT. T_{dev} is set to a new measured RTT. T_{dev} is set to a half of the new measured RTT. This reset operation allows the modified loss differentiation scheme to adapting itself to the change in propagation delay due to vertical handover to new wireless network. Furthermore, the modified loss differentiation is not invoked by three-duplicate ACKs for n rounds. This pause prevents the congestion window from reacting to congestion losses or the number of dropped packets with a set of parameters which have become invalid on vertical handover. The parameters include , T_{min}, T_{avg}, and T_{dev}. In our simulation, we set n to be equal to 1. In our simulation, the values of the parameters measured for the first time after the handover are used to correctly infer the cause of packet losses.

From slow-start state, the state changes into congestion avoidance which grows the congestion window by 1 when the size of congestion window goes higher than the threshold. Three-duplicate ACKs involve the modified LD and DCR. When the third duplicate ACK is received, the LD determines the cause of a packet loss, and the sender with DCR waits for a normal ACK for the duration of one RTT. The waiting duration can be adjusted according to network condition, for example, wireless loss of retransmitted packets and we use the duration used in TCP-DCR [8]. If the sender receives a normal ACK, the state stays in congestion avoidance. Otherwise, the state moves to the fast recovery if the modified LD identifies congestion loss or stays in the congestion avoidance otherwise (wireless loss). In congestion avoidance, the timeout involves F-RTO as described above.

In summary, this proposed scheme II uses the handover detection mechanism from F-RTO to reset the parameters to be used by the modified LD and incorporates the TCP-DCR scheme to improve throughput over 3G network. The LD scheme uses the parameters such as , T_{cur}, T_{avg}, and T_{dev} which already exist in the original TCP. T_{min} is the minimum value of T_{cur}. The DCR uses only T_{avg} to wait for recovery of lost packet at link layer. F-RTO scheme does not use any information. The proposed scheme II needs no extra information other than available in a TCP sender. Thus the proposed scheme II can be implemented easily at the sender-side without involving any change in intermediate routers or receivers.

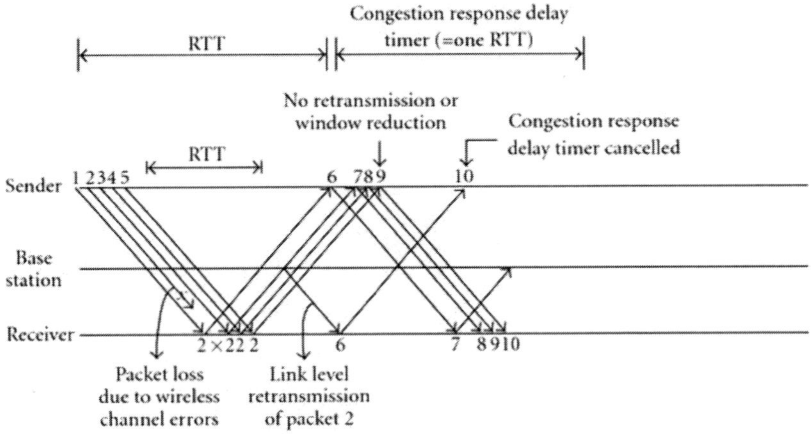

Figure 5. BEHAVIOR OF THE PROPOSED SCHEME II: PACKET-REORDERING DUE TO RETRANSMISSION AT WIRELESS LINK LAYER.

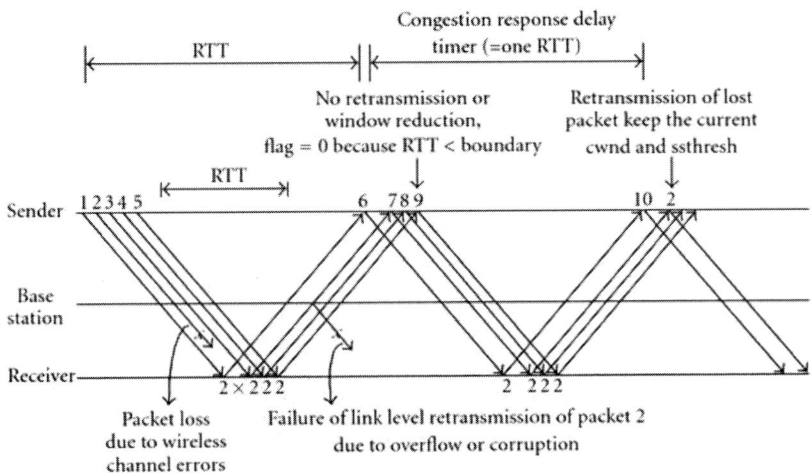

Figure 6. BEHAVIOR OF THE PROPOSED SCHEME II: A WIRELESS LOSS.

4.3. Behavior on Packet Loss

The behaviors of proposed scheme II are shown in Figures 5, 6, and 7. On receipt of the third duplicate ACK, the scheme determines whether the network is congested or not. A flag indicating congestion is set to true if inequality (1) is satisfied, or set to false, otherwise. We recall the equation as follows:

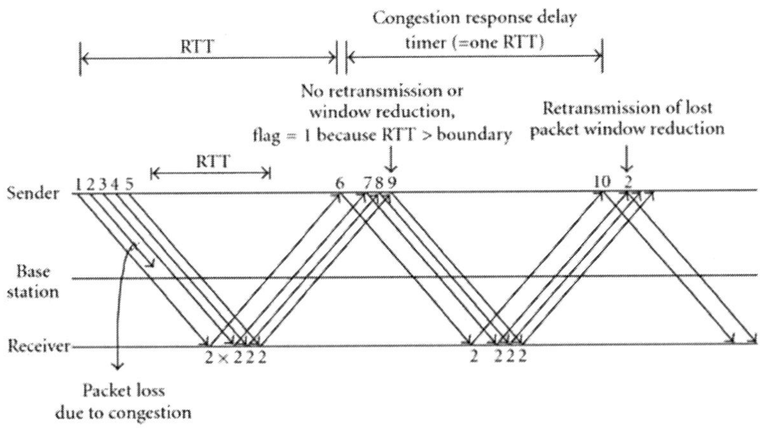

Figure 7. BEHAVIOR OF THE PROPOSED SCHEME II: A CONGESTION LOSS.

$$\frac{T_{\text{cur}} - T_{\text{avg}}}{T_{\text{dev}}} > 2\left(\frac{T_{\text{min}}}{T_{\text{cur}}}\right)^{k} - 1. \quad (3)$$

The scheme then sets the timer to one RTT to wait for link level retransmission of the packets requested in the 3rd duplicate ACK and continues transmitting new packets without reducing the congestion window. When a new ACK is received within one RTT, the sender cancels the timer as shown in Figure 5. This part overlaps with the scheme in TCP-DCR [8].

However, the retransmission procedure is involved when duplicate ACKs are still received after one RTT, and then the decision of retaining or halving the current window is made according to the congestion flag. If it is false, the TCP sender retains the current congestion window as shown in Figure 6. If it is true, the TCP sender halves the congestion window as shown in Figure 7.

The DCR part can handle certain patterns of packet-reordering, but not all possible patterns. For example, packet-reordering of 1-3-2 can be handled since it is a swapping of a packet with the next packet. DCR waits for one RTT after receiving duplicate ACKs and thus can handle such packet-reordering. However, DCR scheme cannot handle the packet-reordering of 1-5-4-2-3 because waiting for one RTT might not be enough for this case.

4.4. Behaviors on Timeout Event

On the timeout event, detailed behaviors are shown in Figures 8 and 9 for a vertical handover and a normal timeout, respectively.

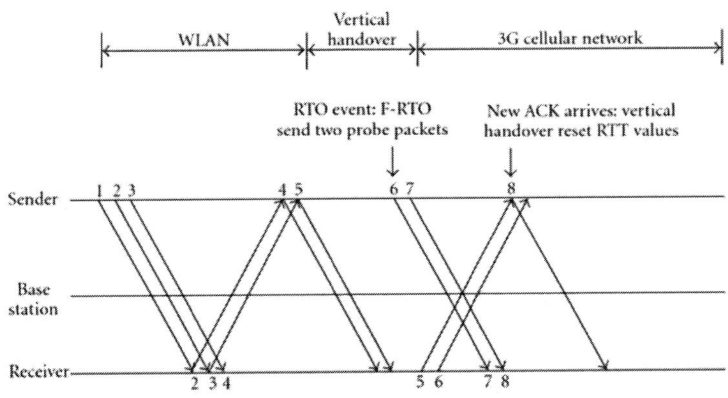

Figure 8. BEHAVIOR ON VERTICAL HANDOVER.

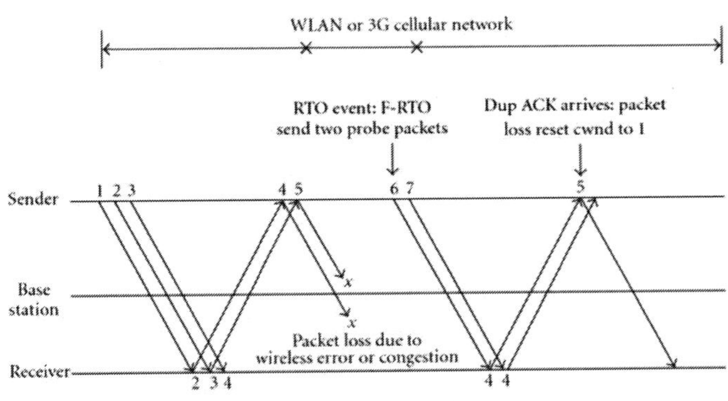

Figure 9. BEHAVIOR ON NORMAL TIMEOUT.

F-RTO scheme is involved to send two probe packets and to wait for ACKs. If a new ACK arrives, the sender regards the timeout event as due to vertical handover (see Figure 8) and resets RTT values (T_{min}, T_{avg}, and T_{dev}) to the initial values. However, if duplicate ACKs arrive, the sender retransmits the lost packets immediately with setting congestion window to 1 (see Figure 9).

5. PERFORMANCE EVALUATION

5.1. Simulation Setup

In this section, we evaluate the TCP throughput performance of TCP-Sack, DCR, DCR-FRTO, proposed scheme I and proposed scheme II through the ns-2 simulations (Version 2.27) [20]. The proposed scheme I is a simple combination of TCP-DCR and LD. The proposed scheme II is introduced in Section 4.2.

The network model is shown in Figure 10. Two scenarios of vertical handover are considered. The first scenario is 3G-to-WLAN handover which denotes handover from 3G to WLAN, and the second one is (W-to-3G) which denotes handover from WLAN to 3G. For each scenario, we also consider two different situations which are noncongestion and congestion. In noncongestion scenarios, the proposed schemes are able to improve TCP throughput on both wireless losses and packet-reordering compared to TCP-SACK and TCP-DCR. Congestion scenarios are more realistic cases with forward cross traffic causing congestion losses over wired paths while wireless packet losses and packet-reordering still happen on wireless paths. The reason for considering congestion scenarios is to test the ability of the proposed schemes to differentiate the congestion loss on handover where parameters experience rapid change. In these congestion scenarios, we mainly focus on the throughput improvement by the proposed scheme II on congestion losses compared to the proposed scheme I as well as known schemes in literature.

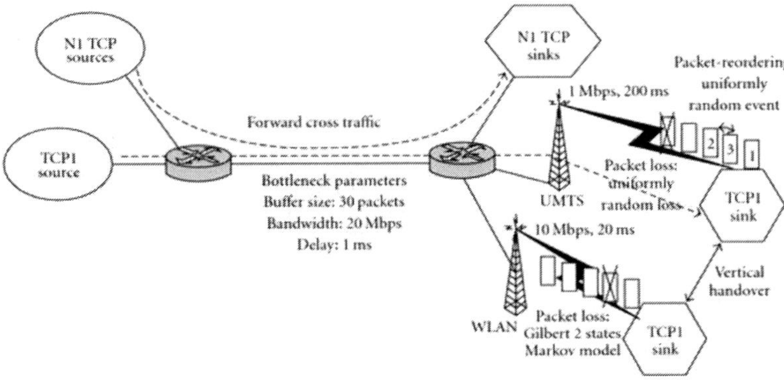

Figure 10. NETWORK MODEL FOR VERTICAL HANDOVER.

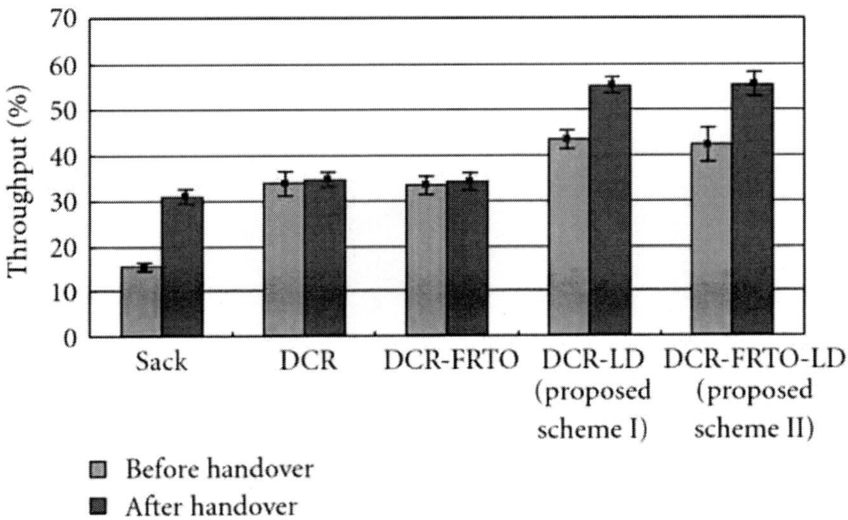

Figure 11. HANDOVER FROM CELLULAR NETWORK TO WLAN.

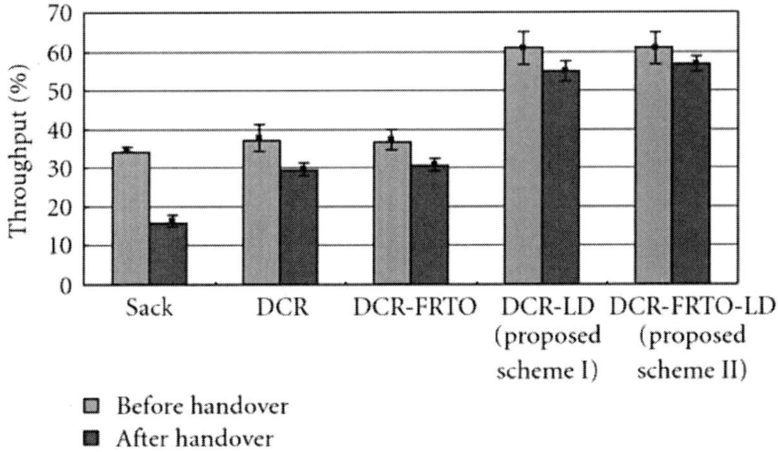

Figure 12. HANDOVER FROM WLAN TO CELLULAR NETWORK.

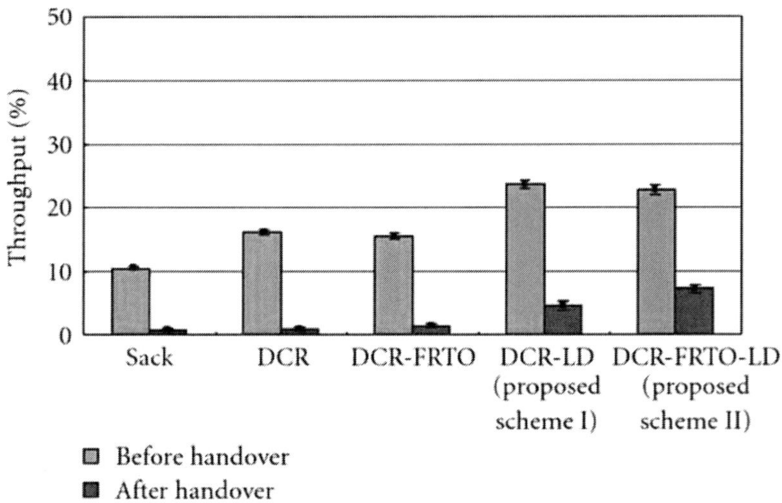

Figure 13. HANDOVER FROM CELLULAR NETWORK TO WLAN UNDER CONGESTION.

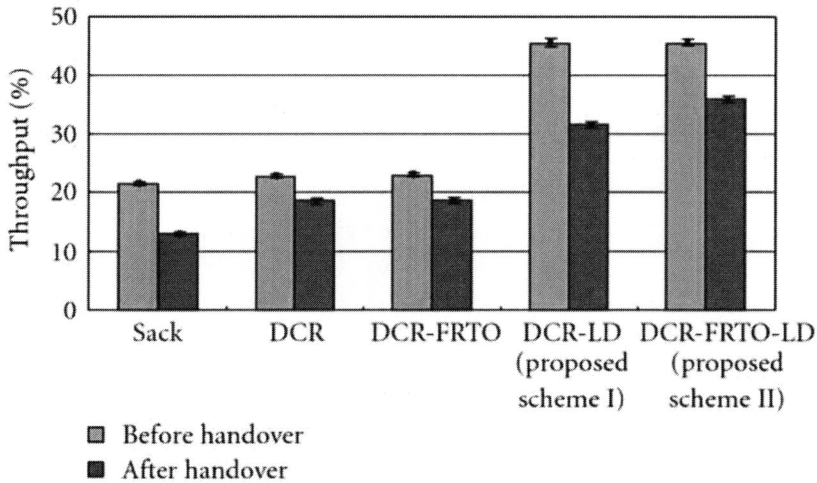

Figure 14. HANDOVER FROM WLAN TO CELLULAR NETWORK UNDER CONGESTION.

In the simulation, we employ a pause event for 30 seconds as a vertical handover from a network to another network. The change in network parameters follows the pause event. The simulation time is equal to 500 seconds. Simulations run 30 times. 95% confidence interval is shown along with the average TCP throughput.

In the simulations, we use the Gilbert two-state Markov chain to model the packet loss process over WLAN [15]. In particular, we assume that $P_{w,WLAN}$ is set equal to 0.1% when the channel is in the Good state and is set equal to 10% when the channel is in the Bad state, deterministically. The duration in the Good state is assumed to be equal to 1 second whereas the duration in the Bad state is assumed to be equal to 100 milliseconds. After the duration, the state transits to the Good state or the Bad state with a probability $P = .5$. The correlated loss model is generally used for modeling the packet loss model of IEEE 802.11 wireless LAN [15].

For the packet loss model over 3G link, we set the packet reordering rate to be as high as 5% and the packet loss rate to be as low as 0.5% with uniform distribution. The packet-reordering can be simply produced by swapping a packet for the next packet. We follow the simulation scenario in TCP-DCR [8] assuming that the underlying mechanism is a simple link-level retransmission scheme, possibly NACK-based, that does not attempt in-order delivery [8, 16].

For the congestion scenario, we inject the forward cross traffic consisting of several TCP connections to cause 1.5% congestion losses rate. This cross traffic also increases the end-to-end delay over the path due to queuing delay. We summarize the configurations in Table 1.

The TCP throughputs are represented as percentages of the maximum bandwidth of the corresponding networks. If a mobile node performs handover from 3G cellular network of 1 Mbps to WLAN of 10 Mbps as in Table 1, 50% TCP throughput before handover is translated to 0.5 Mbps while 40% TCP throughput after handover is translated to 4 Mbps. Please refer to Figures 11–14.

5.2. Throughput Improvement with Vertical Handover under Noncongestion

We evaluate the throughput of a single TCP connection during lifetime including one single handover event between WLAN and cellular network. The throughput of the TCP connection can be affected by wireless packet loss over WLAN and cellular network, packet-reordering over cellular network, and delay spike due to handover process.

Figure 11 shows TCP throughputs for proposed schemes (I and II) against known schemes when a mobile node performs the handover from cellular network to WLAN. Before the handover, the proposed schemes I and II improve the throughput by 26% compared with DCR. This result indicates that the modified loss differentiator improves TCP throughput of the DCR scheme. After the handover to WLAN, the packet-reordering does not happen any more. The only possible loss pattern is wireless loss (noncongestion). The proposed schemes improve the throughput by 60% compared with other schemes (Sack, DCR, and DCR-FRTO).

Figure 12 shows TCP throughputs for proposed schemes (I and II) against known schemes when a mobile node performs the handover from WLAN to cellular network. Before the handover, proposed schemes I and II improve the throughput over WLAN by about 85% compared with Sack, DCR, and DCR-FRTO. The difference between proposed schemes I and II is negligible. Since there are only wireless losses over WLAN (noncongestion), throughputs for the other TCP schemes without loss differentiator are limited. After the handover, the loss pattern changes to coexistence of packet losses and/or packet-reordering. Proposed schemes I and II can achieve the throughput as high as up to 60% of the link bandwidth, and they improve the throughput by about 87% compared with DCR and DCR-FRTO. DCR scheme improves the throughput by 100% compared with Sack, but DCR scheme cannot help halving its congestion window on packet losses in 3G communication system [10].

F-RTO facilitates the TCP sender's detection of the spurious timeout inferred from delay spike without packet losses and prevents the TCP connection from entering the timeout process. When the mobile node moves from WLAN to 3G cellular network, DCR-FRTO and proposed scheme II slightly improves the throughput by 1.2% and 1.8% compared with DCR and proposed scheme I, respectively. However, when the mobile node moves from cellular network to WLAN, the improvement can be negligible. Without F-RTO, the congestion window is shrunk to 1 due to RTO on a vertical handover. But, the high bandwidth of the new fast network can help the congestion window to be recovered to the previous value in a short time. Thus, F-RTO can affect the improvement in throughput only for the case of vertical handover from WLAN to cellular network.

5.3. Throughput Improvement with Vertical Handover under Congestion

In this scenario, we assume that the wired path is congested with congestion packet loss rate of 1.5% due to the forward cross traffic. The proposed schemes improve the throughput by about 44% compared with DCR over cellular network before handover. However, after handover to WLAN, both of proposed schemes I and II improve the throughput by 480% and 457% compared with DCR and DCR-FRTO, respectively. The reason for this great difference is that unlike proposed schemes, the throughputs of previous schemes are suppressed by congestion after handover from a slow link to a fast link.

Figure 13 shows TCP throughputs for proposed schemes (I and II) against known schemes when a mobile node performs the handover from cellular network to WLAN with congestion in the wired part while Figure 14 shows TCP throughputs when a mobile node performs the handover from WLAN to cellular network. After handover to WLAN, throughput is expected to increase due to larger bandwidth over WLAN. Proposed scheme II can increase the throughput from 229 kbps to 734 kbps. Proposed scheme I increases the throughput from 238 kbps to 459 kbps. In these two cases, the modified LD has made the TCP throughput to be higher with differentiation of wireless losses from congestion losses.

In the proposed scheme I, LD regards most of packet losses as wireless losses just after the handover. The threshold of LD is larger than the newly measured RTT because the mean RTT and the deviation remain high immediately after the handover. This makes TCP keep current congestion window even on a congestion loss, which may eventually result in an unnecessary RTO. The RTO holds transmission of normal packets during timeout and reduces its congestion window to 1 after the timeout. Only when the mean RTT and the deviation are adapted to the WLAN, the LD can correctly detect congestion losses.

This is remedied in proposed scheme II. The mean RTT and the deviation are updated on handover to new network. This allows proposed scheme II to adapt to new network faster than the proposed scheme I. Furthermore, the congestion window just after handover is kept to be as same as a value before handover by F-RTO. This implies that the LD in proposed scheme II can detect congestion losses just after handover. The TCP can avoid RTO. This explains why proposed scheme II improves the TCP throughput by 59.9% compared to proposed scheme I over WLAN after the handover. Throughputs for other schemes are reduced by about 50% compared to throughputs before handover.

In the scenario of the handover from WLAN to cellular link over the congested path, the throughput of proposed scheme II in Figure 14 is still higher than other schemes. Especially, the throughput of proposed scheme II is 4.5% higher than proposed scheme I over the cellular network after the handover. This improvement is due to our modification as above mentioned.

The difference between proposed schemes I and II is negligible in noncongestion scenario (Figures 11 and 12). The large difference in WLAN throughput in Figures 13 and 14 can be explained as follows. In our simulation, the forward cross traffic is added on the wired path which has much shorter latency than the wireless paths (3G and WLAN). The added traffic dominates the TCP traffic on WLAN or 3G since it flows only over wired path of high bandwidth (20 Mbps) and very short delay (1 millisecond). Still, the throughput of TCP traffic over WLAN is much larger than the throughput of TCP traffic over the 3G path since the WLAN has much larger bandwidth (10 Mbps) and much shorter delay (20 milliseconds) than 3G path of 1 Mbps bandwidth and 200 milliseconds delay. Thus, before the handover, the suppressed TCP throughput over 3G is much smaller than that over WLAN. Immediately after the handover to WLAN, the TCP throughput is slow to recover in previous schemes (Sack, DCR, DCR-FRTO) and is faster to recover in proposed schemes, but all still remain small due to the limited TCP throughput over 3G before handover as shown in Figure 13. However, the situation is different when handover is performed from WLAN to 3G as shown in Figure 14. The TCP throughput over WLAN before handover is large compared with that over 3G before handover. Thus the TCP throughput over 3G after handover is large whether the schemes are slow to recover (in previous schemes) or fast (proposed schemes). Thus, the big difference in WLAN throughput can be explained by the difference in TCP throughputs over 3G and WLAN under congested condition.

6. CONCLUSION

We modified the TCP congestion control scheme to improve throughput over the vertical handover between two most popular wireless networks, that is, 3G cellular networks and wireless LANs. The proposed scheme is an end-to-end scheme which is easy to implement because it uses information readily available to TCP sender. The modified loss differentiation which employs handover detector scheme from F-RTO and incorporates TCP-DCR is shown to greatly improve the TCP throughput against the coexistence of congestion losses, wireless packet losses, and packet-reordering. When the vertical handover occurs, the proposed scheme is robust to the change in loss pattern over wireless link. Our modification can avoid unnecessary retransmission and reduction of congestion window invoked by spurious timeouts during handover. We find that our modification excels other known schemes (Sack, DCR, FRTO) even when the wired part is congested.

ACKNOWLEDGMENT

This research was supported by Basic Science Research Program through the National Research Foundation of Korea (NRF) funded by the Ministry of Education, Science and Technology (2009-0077261).

AUTHORS' AFFILIATIONS

(1) DMC R&D Center, Samsung Electronics Co., LTD.

(2) Department of Electronic Engineering, Sogang University

REFERENCES

1. *Stemm M, Katz RH: Vertical handoffs in wireless overlay networks. Mobile Networks and Applications 1998, 3(4):335-350. 10.1023/A:1019197320544*
2. *Yosuke M, Takahiro M, Miki Y: TCP congestion control with ACK-pacing for vertical handover. Proceedings of the IEEE Wireless Communications and Networking Conference (WCNC '05), March 2005 3: 1497-1502.*
3. *Huang H, Cai J: Improving TCP performance during soft vertical handoff. Proceedings of the 19th International Conference on Advanced Information Networking and Applications (AINA '05), March 2005 2: 329-332.*
4. *Gou Y, Pearce DAJ, Mitchell PD: A receiver-based vertical handover mechanism for TCP congestion control. IEEE Transactions on Wireless Communications 2006, 5(10):2824-2833.*
5. *Kim S-E, Copeland JA: Interworking between WLANs and 3G networks: TCP challenges. Proceedings of the IEEE Wireless Communications and Networking Conference (WCNC '04), 2004 2: 1252-1257.*
6. *Sardar B, Saha D: A survey of TCP enhancements for last-hop wireless networks. IEEE Communications Surveys & Tutorials 2006, 8(3):20-34*
7. *Lim C-H, Jang J-W: Robust end-to-end loss differentiation scheme for transport control protocol over wired/wireless networks. IET Communications 2008, 2(2):284-291. 10.1049/iet-com:20060172*
8. *Bhandarkar S, Sadry NE, Narasimha-Reddy AL, Vaidya NH: TCP-DCR: a novel protocol for tolerating wireless channel errors. IEEE Transactions on Mobile Computing 2005, 4(5):517-529.*
9. *Sarolahti P, Kojo M, Raatikainen K: F-RTO: an enhanced recovery algorithm for TCP retransmission timeouts. ACM SIGCOMM Computer Communication Review 2003, 33(2):51-63. 10.1145/956981.956987*
10. *Chen J, Leung VCM: Improving end-to-end quality of services in 3G wireless networks by wireless early regulation of realtime flows. Proceedings of the IEEE International Symposium on Personal, Indoor and Mobile Radio Communications (PIMRC '03), September 2003, Beijing, China*
11. *Aguayo D, Bicket J, Biswas S, Judd G, Morris R: Link-level measurements from an 802.11b mesh network. Proceedings of the International*

Conference of the Special Interest Group on Data Communication (SIGCOMM '04), August 2004, Portland, Ore, USA

12. *IEEE 802.11 Part 11 : Wireless LAN medium access control (MAC) and physical layer (PHY) specification: higher-speed physical layer extension in the 2.4 GHz band. IEEE, 1999*
13. *Fantacci R: Queuing analysis of the selective repeat automatic repeat request protocol wireless packet networks. IEEE Transactions on Vehicular Technology 1996, 45(2):258-264. 10.1109/25.492849*
14. *Rossi M, Zorzi M: Analysis and heuristics for the characterization of selective repeat ARQ delay statistics over wireless channels. IEEE Transactions on Vehicular Technology 2003, 52(5):1365-1377. 10.1109/TVT.2003.815980*
15. *Gurtov A, Floyd S: Modeling wireless links for transport protocols. ACM SIGCOMM Computer Communications Review 2004, 34(2):85-96. 10.1145/997150.997159*
16. *TCP-DCR source codes for ns-2 simulations http://dropzone.tamu.edu/~sumitha/research.html*
17. *IEEE Std P802.21/D02.00 IEEE Standard for Local and Metropolitan Area Networks: Media Independent Handoff Services, September 2006*
18. *Choi YS, Lee W, Cho YZ: TCP with explicit handoff notification for a seamless vertical handover. Proceedings of International Conference on Information Networking (ICOIN '08), 2008, Lecture Notes in Computer Science 5200: 831-840.*
19. *Gurtov A, Korhonen J: Effect of vertical handovers on performance of TCP-friendly rate control. ACM SIGMOBILE Mobile Computing and Communications Review 2004, 8(3):73-87. 10.1145/1031483.1031494*
20. *NS-2 network simulator (ver 2). LBL*

7

CHAPTER

THE MAKING OF A SUSTAINABLE WIRELESS CITY? MAPPING PUBLIC WI-FI ACCESS IN SHANGHAI

Mingfeng Wang [1], Felix Haifeng Liao [2,*], Juan Lin [1], Li Huang [2], Chengcheng Gu [1] and Yehua Dennis Wei [3]

ABSTRACT

In the context of the global information economy, ready access to the Internet is critical to a city's competitiveness, which has prompted a number of cities to launch plans to establish wireless networks. Most literature on the development of wireless cities focuses on cities in Western countries, and few have discussed how Chinese cities have adopted wireless technologies in their urban infrastructure development efforts. This paper examines recent development and spatial distribution of public Wi-Fi access in Shanghai, a leading business hub in China. We mapped Wi-Fi hotspots through the government sponsored "i-Shanghai" project and China Mobile Communications Corporation (CMCC). We find that while telecommunication providers have been proactively deploying WLAN (wireless local area network, a proxy of public Wi-Fi or wireless access) hotspots in Shanghai, neither government sponsored WLAN hotspots nor facilities established by CMCC could cover the old traditional neighborhoods in the central city and sub-districts in remote rural areas. We also address the development of a more sustainable wireless city in Shanghai with a particular focus on digital divide and social equity issues.

Keywords: Wi-Fi; wireless cities; GIS spatial analysis; digital divide; China

1. INTRODUCTION

Wireless communication technologies have emerged as a major force underlying the recent development and change of the global economy [1]. Under such notions as digital cities [2], intelligent cities [3], mobile cities [4], wireless cities [5], ubiquitous cities [6], and smart cities [7,8], new planning strategies that emphasize the adoption and adaptation of information and communication technologies (ICTs) have attracted considerable attention from scholars and policy makers. In the context of the global information economy, the usage of wireless technologies has become a key indicator of the competitiveness of a city [5,9,10,11]. A number of cities, such as Singapore and Taipei in Asia, Philadelphia, San Francisco and Boston in the United States, and Perth in Australia, have either expressed an intention to establish a wireless network or launched specific plans to develop wireless cities [5,12,13]. While most literature on wireless cities and the digital divide focuses on cities in Western countries, few have discussed how Chinese cities have adopted wireless technologies in infrastructure development, and even fewer have addressed the issue of digital divide in China's urban landscapes.

By the end of 2014, China's mobile Internet users had reached 649 million, as compared to 22.5 million in 2000, accounting for 19% of the total number in the world [14], among which the number of smart phone users increased to 557 million by the end of 2014 [15]. Internet development, especially a dramatic increase in the number of smart phone users, has resulted in a booming demand for wireless access in Chinese cities [16]. In collaboration with major telecommunication operators, including China Mobile Communications Corporation (CMCC), China Unicom, and China Telecom, local governments in Chinese cities have launched plans for public Wi-Fi network access and the development of "wireless" cities. For example, since 2012, the Shanghai Municipal Government has launched a project named "i-Shanghai" (or loving Shanghai). The project aims to deploy over 4000 Wi-Fi hotspots by 2020, making Shanghai a large-scale wireless city.

This paper examines the development of public Wi-Fi access in Shanghai and compares the WLAN hotspots established by CMCC and those sponsored by the Shanghai Municipal Government through the "i-Shanghai" project. Notably, the total number of Internet users in Shanghai has increased by 500% during 2000–2014, among which the share of Internet users through smart phone or other mobile devices has increased from

64% in 2009 to 79% in 2013 [17,18]. We analyze the spatial distribution of Wi-Fi access at multiple scales (i.e., district and sub-district levels) and in different areas (e.g., central city and suburban areas). By applying exploratory spatial data analysis (ESDA) methods, such as concentric analysis and spatial hot spot analysis, we underscore the proactive role played by telecommunication companies in shaping the Wi-Fi geographies in Chinese cities. The remainder of the paper is organized as follows: the next section briefly reviews the concept of wireless city and related literature on wireless city development and the digital divide. This is followed by the spatial distribution analysis of public Wi-Fi access in Shanghai provided by the "i-Shanghai" project and CMCC. The last section summarizes the major findings and discusses the characteristics of Shanghai's pathway towards a sustainable wireless city.

2. BACKGROUND AND LITERATURE REVIEW

Wireless cities are essentially cities completely covered by high-speed broadband and public Wi-Fi access, where the Internet can be openly accessed and used by their citizens. Wireless technologies are an important part of infrastructure development through which the efficiency and equity of governmental service can be improved [5]. Wireless network coverage is also a key step to extend fiber broadband to the public, and the access to Wi-Fi, and more broadly the Internet, is regarded as "the city's fifth major infrastructure" next to water, electricity, gas, and roads [19,20,21]. The impact of wireless city development is significant in many aspects, such as providing mobility values [22], changing travel behaviors [23], altering the perception of community and forming social networks in urban space [24,25,26]. Previous literature found that in wireless cities, transaction costs can be reduced [27] and social well-being of citizens is better off through closer social interactions [22,28,29].

Given the benefits of developing wireless networks or ready access to the Internet, a great number of cities are deploying or have plans to establish wireless broadband networks [30]. Previous literature has also documented a number of factors that determine the locations of public Wi-Fi access mostly in Western cities. Customer choice is regarded as a basic driving force behind the development of the wireless broadband market, and the layout of wireless facilities is obviously influenced by local social and economic factors [31]. Oyana [32] studied the distribution of wireless facilities in Southern Illinois, USA and identified three main factors that influence the distribution of wireless facilities: the higher educational population rate, age-specific group, and average family income. Driskella and Wang [33] examined the spatial layout of Wi-Fi access in Louisiana, USA, suggesting that the determinants of public Wi-Fi hotspot location are residents' socio-economic disadvantages in neighborhoods and their household characteristics. Grubesic and Murray [31] analyzed the distribution of Wi-Fi hotspots in four neighborhood communities in Cincinnati, USA, and reported that the degree of network access in each community was closely related to social, economic, demographic, and spatial factors, but evident differences exist between rich and poor blocks' Wi-Fi access; the network access in the inner city is concentrated in areas around commercial and office buildings.

Urban planning and government policies also play an important role in the making of a wireless city [34]. For example, through a comparative study of Singapore and Taipei, Hu et al. [5] suggested that the development of the wireless city in Taipei is mainly driven by top-down government policy while such development depends more on the market-driven participation of customers in Singapore. The development of wireless city is also driven by the innovation and widespread use of smartphone technologies. In a case study of Salt Lake City, USA, Torrens [35] identified that Wi-Fi access has provided a possible solution to "last-mile" problems in the city, and Evans-Cowley [36] also addressed the influences of mobile phones on urban life.

Given the nature of Internet access as a public utility, the role of government in providing Wi-Fi infrastructure is of particular concern in the literature [30,37,38]. Ballon et al. [30] reviewed the public-private partnership in shaping wireless city networks in both EU and the U.S. and found that public authorities such as governments will strive for the optimal trade-off between minimizing their inputs for directly being involved in the establishment and operation of wireless city networks and maximizing the leverage for the purpose of specific policy goals such as narrowing the digital divide and so on. Girth [37] explored the variation in approaches and examined the structural factors that give rise to public-private partnerships. Some literature also have examined the role of the Asian government, China especially, in telecommunication industry development and infrastructure construction. For instance, Soh and Yu [39] analyzed the development of 3G networks in China and explained how regulatory impacts from government and domestic and foreign network operators are interdependent to each other.

Despite the positive impacts of the wireless city development, the spatial and social inequalities of information technology and its access has also drawn a lot of attention [40,41], derived from the traditional thoughts of urban sociology [42], with a particular focus on the digital divide. Moreover, this type of restructuring is often associated with a new round of social differentiation in different urban spaces [31,43]. From this perspective, the term digital divide, which refers to the fact that certain segments of the population and specific social groups may be excluded from access to the Internet, has gained considerable scholarly attention [5,7,38,44,45]. As Castells [46] pointed out, the gap between network haves and have-nots increases sources of spatial and social inequality. Combined with the historical socioeconomic divides in the city, the strong commercial bias of Internet access may strengthen an uneven geography of public Wi-Fi access [47], which can be manifested at different scales, from global to regional ones, and to neighborhoods and communities [43,48,49,50]. Popularized by Gray Andrew Pole, a New York Times journalist, the notion of digital divide has called for more attention in regards to equitable access to public Wi-Fi and the uneven distribution of access to the Internet within cities [35,48]. For example, Prieger [50] investigated the gaps in broadband usage for minorities and found that fewer fixed broadband options were available to Blacks and Hispanics, but they tend to have more mobile broadband providers available. These digital disparities are closely associated with social polarization and inequalities, imposing new challenges for the sustainability of families, communities, and cities in the context of the new information economy [51].

In short, despite an emerging body of literature on wireless cities and digital divide in Western countries, Chinese cities have largely escaped from scholarly attention and few work has been done to address potential digital divide in the course of the Chinese wireless city development [52]. Specifically, to the best of our knowledge, most of the previous literature on the digital divide or Internet development in China has focused on disparities or spatial distribution at the provincial, city or regional levels [53,54], the development of wireless city and digital divide at the intra-city level have rarely been researched.

3. DATA AND STUDY REGION

The data used in this study was mainly obtained from the official websites of Shanghai CMCC [55] and the "i-Shanghai" project [56]. The raw data contains the attributes of each WLAN hotspot established by CMCC and the "i-Shanghai" project, including the name, address, type, located sub-district, and also the same attributes of Wi-Fi hotspots. WLAN is one of the most widely used broadband wireless technologies that provides public Wi-Fi access in China. By July 2013, the number of WLAN hotspots provided by

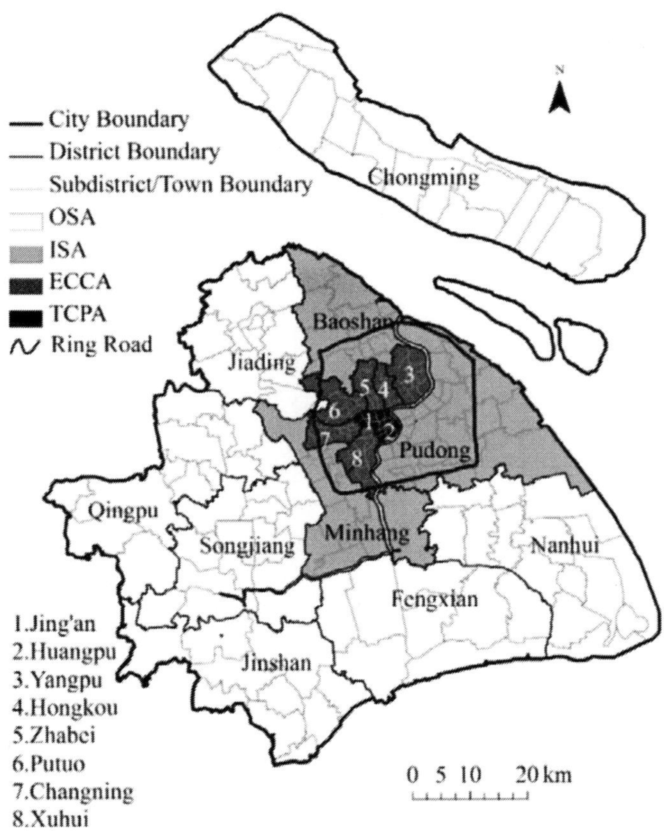

Figure 1. Administrative division and geographic structure of Shanghai. Note: TCPA = traditional city proper area; ECCA = expanded central city area; ISA = inner-suburban area; OSA = outer-suburban area.

Shanghai CMCC had increased to 6800, as compared to less than 703 in 2009. In this study, the locations of all the CMCC and "i-Shanghai" WLAN hotspots in 2013 were geocoded. In addition, as shown in Figure 1, our study area included 18 districts and counties: Huangpu, Jingan, Xuhui, Changning, Putuo, Zhabei, Hongku, Yangpu, Pudong, Minhang, Baoshan, Songjiang, Jiading, Qingpu, Jinshan, Fengxian, Nanhui and Chongming, and 208 sub-districts/towns, which are the smallest administrative unit in China. Following Wei et al. [57], we divided the 18 districts and 208 sub-districts into four areas: traditional city property area (TCPA, which is composed of Huangpu and Jingan), expanded central city area (ECCA, which includes Xuhui, Changning, Putuo, Zhabei, Hongkou and Yangpu), inner-suburban area (ISA, which consists of Pudong, Minhang and Boshan), and outer suburban area (OSA, which includes Chongming, Jiading, Songjiang, Qingpu, Jinshan, Fengxian and Nanhui).

As the largest business hub and financial center in China, Shanghai has played a leading role in developing wireless cities in China. As illustrated in Figure 2, the Internet penetration, as measured by the total number of Internet users divided by total population, increased from less than 30% before 2005 to over 70% in 2013. The number of Internet users through smartphones in Shanghai also increased from 7.47 million to 12.6 million during 2009–2013, and its shares of Internet users rose from 63.8% to 78.5% [17,58]. On par with Beijing, Shanghai is currently one of the two Chinese cities with the highest rates of Internet penetration. In fact, as early as the 1990s, Internet development was one of the items that were prioritized in the agenda of Shanghai municipal government. The theme of the 2010 Shanghai World Expo, "Better City, Better Life", reiterated the importance of developing Shanghai into a wireless city. In 2011, the Shanghai Municipal Government issued "The Shanghai's Promotion Plan for the Development of Smart City (2011–2013)", which focused on the construction of wireless city. In collaboration with major corporations in the Chinese telecommunication industry, especially CMCC, one of the goals in the plan was to build a wireless broadband network that covers over 80% of the city, with the speed of Internet of 20 Mbps.

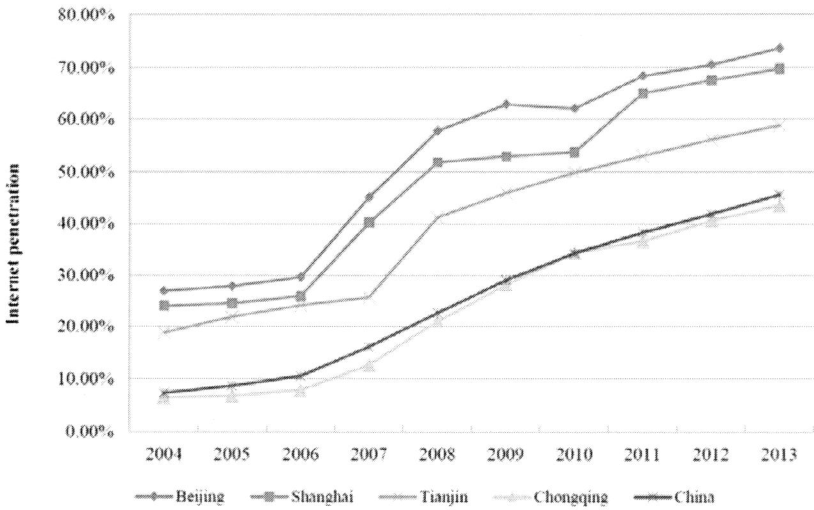

Figure 2. Internet penetration in Shanghai, as compared to China and other centrally administrated municipalities; Source: China Internet Network Information Center, 2004–2013.

4. MAPPING PUBLIC WI-FI ACCESS IN SHANGHAI

This section comprehensively analyzes the geographies of Wi-Fi access in Shanghai, with an emphasis on those hotspots provided by the Shanghai Municipality Government through the "i-Shanghai" project and those established by the CMCC. CMCC has contributed significantly to the wireless city development in Shanghai. In 2013, CMCC had established more than 6000 hotspots, and the majority of these Wi-Fi hotspots are not free to access. The "i-Shanghai" project was launched in 2012, and by the end of 2013, 450 WLAN hotspots were deployed to provide free Wi-Fi access to the public. As these WLAN hotspots are free to access, a lot of benefits have been brought to those who could not afford Internet access, although the free access could only last for 2 hours.

Figure 3 describes different locations of CMCC and "i-Shanghai" WLAN hotspots in Shanghai, indicated by the type of these WLAN hotspots. CMCC WLAN hotspots are more likely to be located in commercial districts and office buildings, which accounted for 26% and 31% of all CMCC hotspots, respectively, in 2013. In comparison with CMCC hotspots, the majority of hotspots provided by the "i-Shanghai" project were placed in those areas associated with governmental and public services, especially places of interest, parks, exhibition and sports centers, and hospitals (Figure 3).

Geographically, both CMCC and "i-Shanghai" hotspots were concentrated in the areas within the outer Ring Road in Shanghai or 15 kilometer to the people's square or the city center (Figure 1). In general, "i-Shanghai" Wi-Fi hotspots were more likely to locate close to the city center (Figure 4b), and the CMCC Wi-Fi hotspots spread out more extensively to suburban districts. It should be noted that, in 2013, only eight out of 206 sub-districts in Shanghai had no CMCC WLAN hotspots and most of these sub-districts/towns were located in outer suburbs like Chongming Island, Jinshan, and Nanhui (Figure 4a). Regardless of CMCC or "i-Shanghai" WLAN hotspots, a few sub-districts/towns with a very small number of WLAN hotspots, say less than five, were located in the traditional city area of Shanghai characterized by geographical concentrations of work force communities and those neighborhoods in remote rural areas (Figure 4).

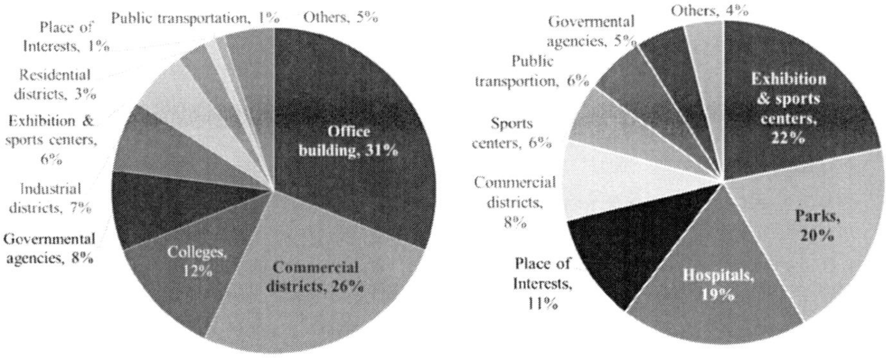

(a) (b)

Figure 3. Types of CMCC and "i-Shanghai" WLAN hotspots, 2013. Source: CMCC and "i-Shanghai". (a) CMCC; (b) "i-Shanghai".

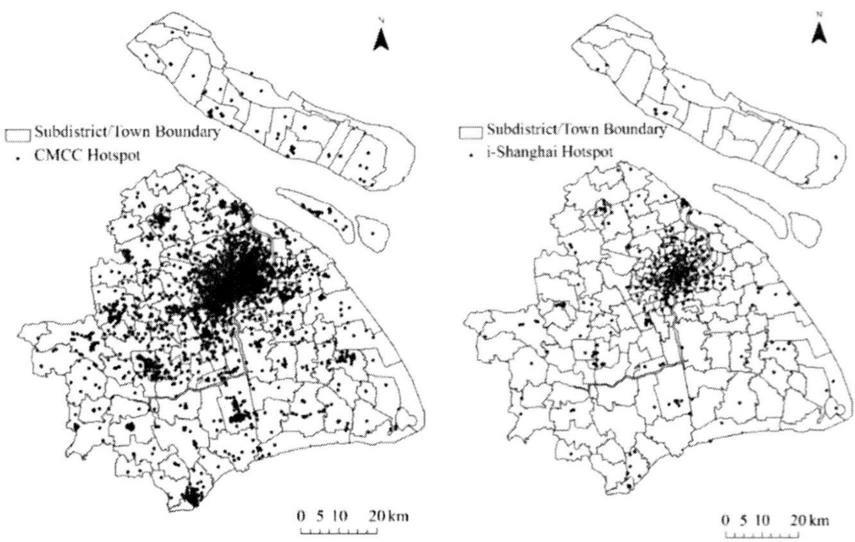

(a) (b)

Figure 4. Spatial Distribution of CMCC and "i-Shanghai" WLAN hotspots, 2013. (a) CMCC; (b) "i-Shanghai".

Table 1 presents the top ten sub-districts/towns with the largest number of CMCC and "i-Shanghai" WLAN hotspots. Some interesting findings emerge: first, in comparison with the list of "i-Shanghai" hotspots, more sub-districts in the suburban areas were found as the top ranking sub-district/towns with the largest number of CMCC WLAN hotspots. Seven out of the top-ten ranking sub-districts were in suburban areas, such as Fangsong in Songjiang district and Jiangchuan Road in Minhang district. Second, the total number of CMCC WLAN hotspots in most of the top-ranking sub districts/towns were installed around a number of office buildings and higher education and R & D institutes, such as Xujiahui in Xuhui district and Zhangjiang in Pudong district. By contrast, the top two sub-districts with the largest number of the "i-Shanghai" hotspots, i.e., Lujiazui and Huapu, were located in the inner-suburban area (ISA). Most of these places are important places of interest and locations where governmental buildings are concentrated. Overall, sub-districts with agglomerations of higher education institutes and R & D centers tend to have more CMCC WLAN hotspots, but the "i-Shanghai" hotspots tend to agglomerate in those landmarks in Shanghai, such as the Bund, Huangpu, and Lujiazui financial district, etc. (Table 1).

Table 1. Ranking of China Mobile Communications Corporation (CMCC) and "i-Shanghai" WLAN (wireless local area network) hotspots in Shanghai's sub-districts/towns in 2013 (Top 10).

CMCC				"i-Shanghai"			
Rank	**Subdistrict**	**Location**	**N**	**Rank**	**Subdistrict**	**Location**	**N**
1	Fangsong	OSA	395	1	Lujiazui	ISA	16
2	Nanqiao	OSA	171	2	Huangpu	ISA	15
3	Huamu	ECCA	158	3	The Bund	TCPA	11
4	Xiayang	OSA	131	4	Chuansha	ISA	10
5	Zhangjiang	ISA	124	5	Nanqiao	OSA	9
6	Jiangchuan	ISA	123	-	Huaihai Zhong Road	TCPA	9
7	The Bund	TCPA	120	7	Jiangsu Road	ECCA	8
8	Xujiahui	ECCA	113	-	Ou-yang Road	ECCA	8
9	Huinan	ISA	102	-	Youyi Road	ISA	8
10	Zhelin	OSA	100		East Nanjing Road	TCPA	7
				10	Fenglin Road	ECCA	7
					Tianping Road	ECCA	7
					Zhoujiadu	ISA	7

TCPA = traditional city proper area; ECCA = expanded central city area; ISA = inner-suburban area; OSA = outer-suburban area.

In order to further explore the Wi-Fi geographies in Shanghai represented by the spatial distribution of CMCC and "i-Shanghai" WLAN hotspots, concentric analysis based on a 5-km radius interval and ESDA methods were used, following Wang and Gu [52]. Results of concentric analysis demonstrate that the share of CMCC WLAN hotspots within 5 kilometers from the city center (Figure 5), which is defined as the location of People's Square, was less than 20%, while its counterpart under the "i-Shanghai" project was over 35%. There were also nearly 40% of the CMCC hotspots located 20 kilometers or further from the center (i.e., outer suburbs), whereas the share of "i-Shanghai" hotspots was just above 25% in the same region (Figure 5).

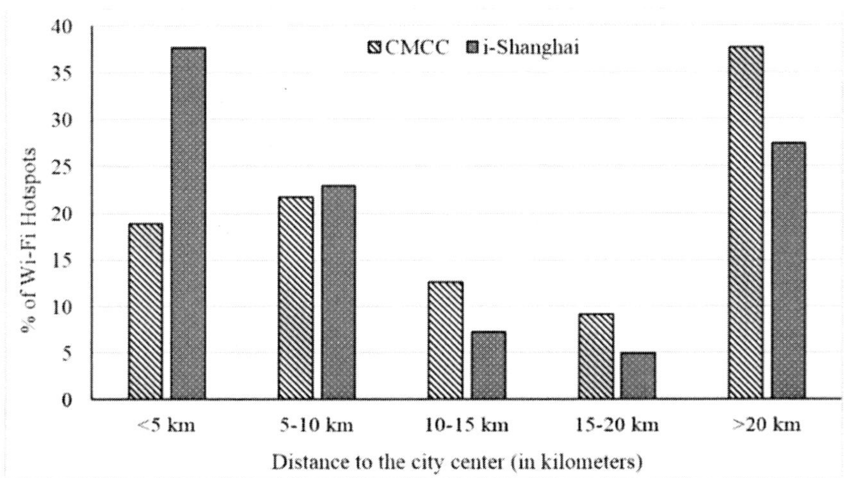

Figure 5. Spatial distribution of "i-Shanghai" and CMCC hotspots based on concentric analysis.

Spatial hot spot analysis (Getis-Ord Gi*), which measures the spatial association of a variable and identifies the characteristics of patterns [59], was used to further detect the clusters of WLAN hotspots at the sub-district/town level. This method detects the statistically significant spatial clusters of high values, or sub-districts/towns with large number of hotspots in this study, and identifies them as "hot spots", and the statistically significant spatial clusters of low values, or sub-districts/towns with little number of hotspots, and identifies them as "cold spots". The Getis-Ord Gi* statistic is widely used in analyzing the distribution of geographically referenced data. The Gi* of each unit i could be calculated as:

$$G_i^* = \frac{\sum_{j=1}^{n} w_{i,j}x_j - \overline{X}\sum_{j=1}^{n} w_{i,j}}{S\sqrt{\frac{\left[n\sum_{j=1}^{n} w_{i,j}^2 - \left(\sum_{j=1}^{n} w_{i,j}\right)^2\right]}{n-1}}}; \quad (1)$$

where x_j is the number of hotspots in administrative unit j, $w_{i,j}$ is the spatial weight matrix, n is the number of administrative units, and S is the standard deviation of the observations. The results of hot spot analysis is essentially a z-score for each spatial unit, measuring the spatial clustering of high/low values. A statistically significant (e.g., $p < 0.05$) positive z-score implies the presence of a hot spot. By contrast, a statistically significant (e.g., $p < 0.05$) negative z-score implies the presence of a cold spot [59,60].

As shown in Figure 6b, a typical "core-periphery" structure of WLAN hotspots was found with respect to "i-Shanghai" Wi-Fi hotspots. The hot spots within the urban area (inside of the Shanghai Outer Ring Expressway) constituted the core area, indicating the clustering of "i-Shanghai" WLAN hotspots. These sub-districts/towns are mostly important commercial districts in Shanghai, such as Xujiahui, the Bund, and Lujiazui financial district. In contrast to hot spots, the cold spots were located in remote suburban sub-districts/towns, favoring less developed districts, such as Chongming Island. In comparison with the spatial pattern of "i-Shanghai" hotspots, the spatial distribution of CMCC Wi-Fi hotspots displayed a bi-center distribution. The downtown area (within five kilometers from the People's Square) remained as the core area, and what is more interesting is that another core area has emerged in the suburbs to the southwestern part of Shanghai. The new core area is where new development zones and new higher education institutes are located, including Songjiang New City, Qingpu New city, Jiangchuan subdistrict, and Nanqiao New Town. A lot of sub-districts are characterized by geographical concentrations of higher education institutes such as Songjiang University town in Songjiang New Town, Haiwan University town in Tuolin town, and Shanghai Jiao Tong University and Shanghai University of Electric Power in Jiangchuan sub-district.

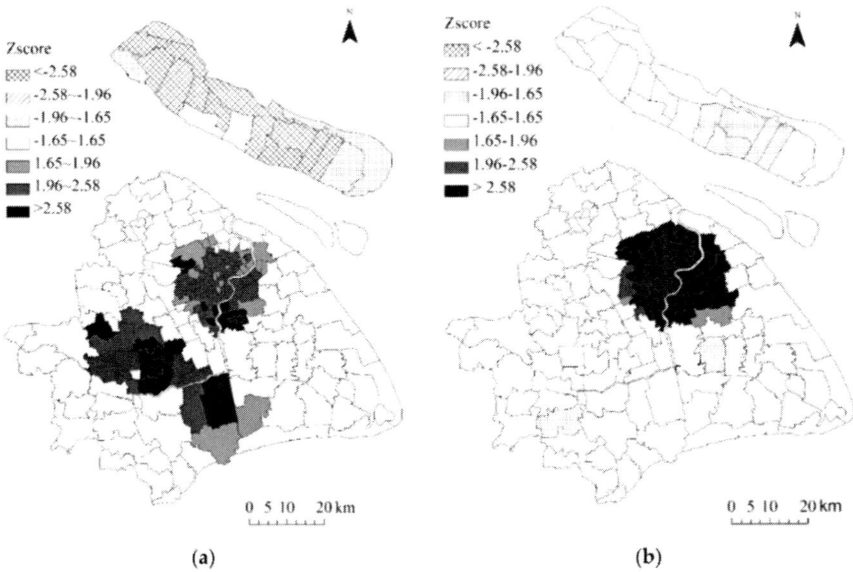

Figure 6. Hotspot analysis of CMCC and "i-Shanghai" hotspots in Shanghai, 2013. (a) CMCC; (b) "i-Shanghai".

In comparison with the spatial layout of "i-Shanghai" hotspots, CMCC tends to be more proactive in deploying WLAN hotspots. As shown in Figure 6a, the bi-center distribution of hot spots of CMCC Wi-Fi hotspots is greatly driven by a rapid process of suburbanization in Shanghai in recent years [61,62]. In this process, large-scale industrial enterprises extend to suburban areas first, followed by the growth of population and new housing development [63]. In addition, given the skyrocketing land price and restricted land supply in the downtown area, many business institutions and government agencies have recently moved to the suburbs in Shanghai. As a result, the rapid population growth and increased demand for Wi-Fi access have prompted CMCC to develop new cyber infrastructure like WLAN hotspots in these sub-districts (Figure 6a).

5. MICRO-SCALE ANALYSIS OF THREE SUB-DISTRICTS

The mapping above has revealed the spatial distribution of WLAN hotspots or public Wi-Fi access in Shanghai. Contrasting Wi-Fi geographies shaped by CMCC and "i-Shanghai" WLAN hotspots were found, which highlight the proactive role of telecommunication operators in Shanghai's wireless city development. The findings motivate the concern regarding how key stakeholders, such as government, telecommunication operators, and communities, can play a role in Shanghai's wireless city development. The following section will focus on three specific sub-districts, including Nanqiao New Town, Songjiang University Town, and Pengpu Xincun, given their unique features regarding the wireless development at the community level (Figure 7).

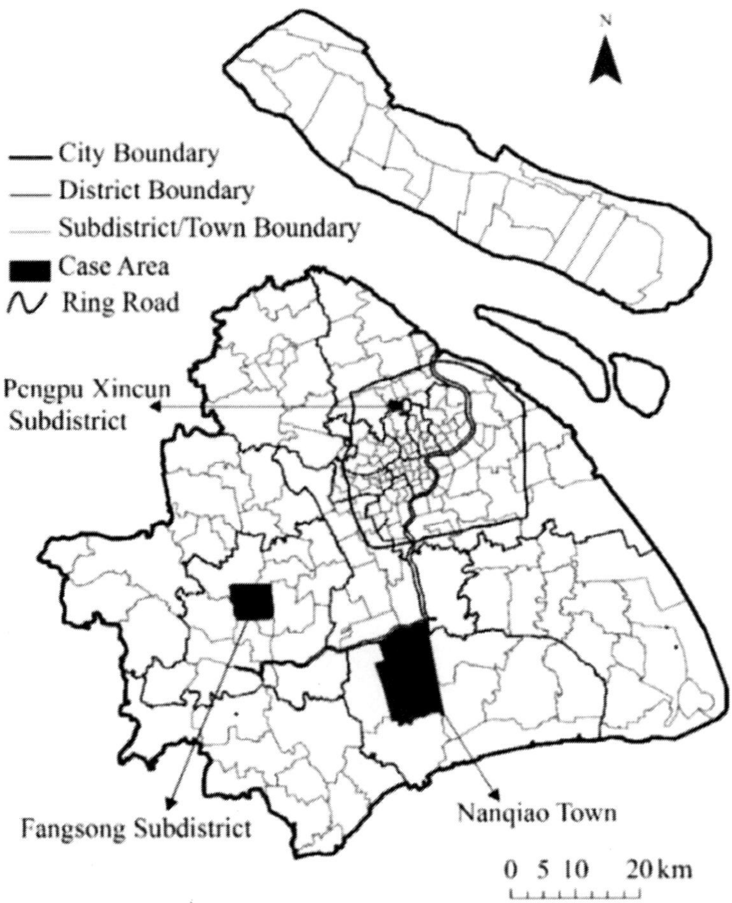

Figure 7. Locations of case studies.

5.1. Nanqiao New Town

Nanqiao is one of the top-ten ranking sub-districts that has experienced a rapid wireless development driven by the deployment of both CMCC and "i-Shanghai" WLAN hotspots (Table 1). In 2013, the number of CMCC hotspots increased to approximately 170, ranking as the second largest one in Shanghai. As shown in Figure 8a, there have been five clusters of CMCC hotspots: 1. Fengpu Commercial Street, especially those within the Fengpu Mansion, which is a commercial office building; 2. Bailian shopping mall; 3. the historical community (old town); 4. Government of the Fengxian District; 5. Nanqiao Township Government. In contrast to the spatial layout of CMCC WLAN hotspots, the distribution of public Wi-Fi access provided by "i-Shanghai" was centered on government agencies, including the Fengxian District Conference Center and Nanqiao Township Government, and they were also located around large-scale businesses companies or the Shopping malls, namely RT-MART and LOTUS (Figure 8b).

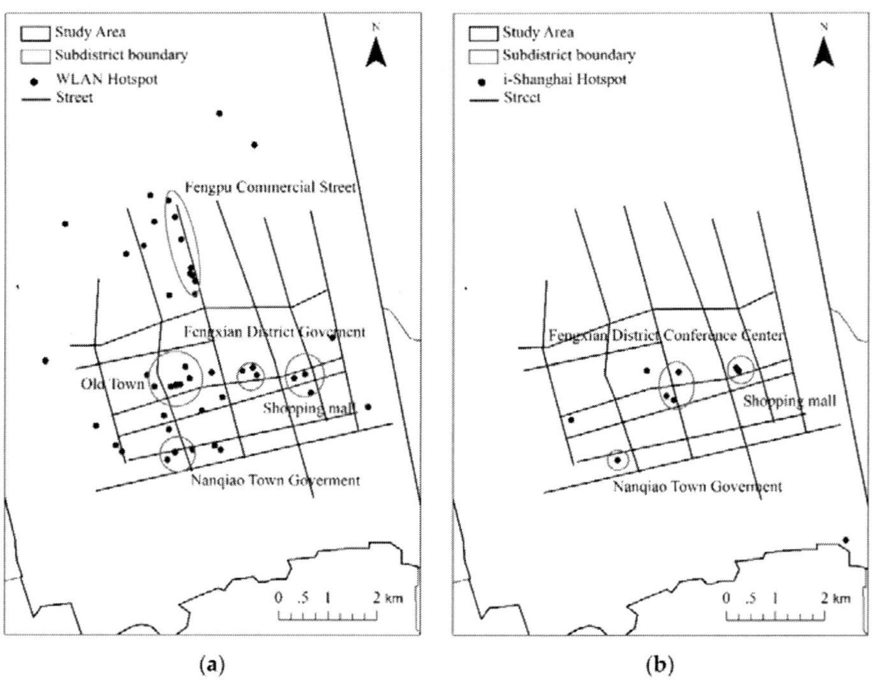

Figure 8. Spatial distribution of CMCC and "i-Shanghai" hotspots in Nanqiao New Town, 2013. (a) CMCC; (b) "i-Shanghai".

In the case of Nanqiao, the government policy and urban planning have played an important role in the expansion of public wireless access. Nanqiao New Town is the political, economic, and cultural center in Fengxian district of Shanghai, and it is also planned to be a core area providing comprehensive services for the north coast of Hangzhou Bay. As one of the three major new towns in Shanghai, the development of Nanqiao was emphasized in Shanghai's 12th Five-Year Plan [64]. Based on its development plan, new ICTs or wireless technologies were the key instrument to integrate various functions within this sub-district. These government investments aimed to provide wireless access especially in the eastern part of the Nanqiao New Town (Figure 8b). It was also designed to provide favorable development conditions for enterprises and high quality living for residents, and to promote a more competitive local business environment, with great potential demand for Internet access.

5.2. Songjiang University Town

Fangsong sub-district in Songjiang is another typical case in terms of wireless development. In comparison with Nanqiao New Town, Fangsong sub-district has achieved a rapid growth of CMCC WLAN hotspots. There have been 395 CMCC hotspots in this area whereas the number of "i-Shanghai" hotspots was four at the end of 2013. These hotspots have provided public Wi-Fi access to thousands of people especially college students. Different from Nanqiao, the wireless development was heavily driven by the establishment of Songjiang University Town. Among the 395 CMCC WLAN hotspots in 2013, about 320 were located within Songjiang University Town. In this sub-district, there are a large number of college students and faculty in the seven higher education institutions, including Shanghai International Studies University, University of International Business and Economics, Shanghai Lixin University of Commerce, Donghua University, East China University of Political Science and Law, and Shanghai University of Engineering Science. The case of Songjiang University Town marks a more proactive role of telecommunication company, i.e., CMCC, in recent wireless development in Shanghai, by particularly tracing the new demand from college students in university towns [65].

5.3. Pengpu Xincun Sub-District

Pengpu Xincun sub-district serves as a case where specific communities have been overlooked in the course of wireless city development in Shanghai. As shown in Figure 7, Pengpu is located in the downtown area. However, there has been very limited public Wi-Fi access either through

CMCC WLAN hotspots or the Wi-Fi access provided by the "i-Shanghai" project. In 2013, only five CMCC WLAN hotspots, as compared to over one hundred CMCC WLAN hotspots in other nearby sub-districts in the downtown of Shanghai. Similarly, there was even no "i-Shanghai" WLAN hotspot located in this sub-district. The lagging development of wireless access in Pengpu has its own historical background. The dwelling houses in the sub-district were built for low-paid workers in previous state owned enterprises. Like other shanty towns in Asian cities [66], most buildings in Pengpu were characterized by old and small living spaces, even though the residential population in Pengpu was relatively large, about 160,000 residents in 2013. The development of Pengpu community has slowed down in recent years because of the limited space, and the unfavorable socio-economic conditions in the community have resulted in the backward wireless development in these areas. In short, the case of Pengpu reflects that even though the WLAN hotspots have been widely built up due to the efforts made by either telecommunication companies or the government, there have been many new "blind spots" in Shanghai's communities where the gap between information haves and have-nots is still evident [67].

6. POLICY IMPLICATIONS

Our mapping of public Wi-Fi access in Shanghai has important policy implications in developing wireless cities and addressing the digital divide issue in the context of China. First, as shown in this study, the planning in China is more focused on physical dimensions and deemphasizes issues related to social equity [68], as illustrated by the digital disparities at the community and neighborhood levels in this study. More efforts should be made to address the social dimension of wireless city development in Shanghai and other Chinese cities. Second, Chinese cities are transitional cities. Local governments and market forces are collectively shaping the Wi-Fi geographies especially through the establishment of new development zones and towns in their suburban areas [69]. Hence, how to develop a mutually beneficial collaborative relationship between the government and market participants should be underscored in the future infrastructure planning. Third, although the study has been focused on the spatial distribution of public Wi-Fi access in Shanghai, the development of wireless cities should address not only the construction of hardware facilities, but also the education and supporting system that improves access to the Internet. Therefore, not only the quantity of wireless access, but also the quality of public access to the digital world, deserves attention from policy makers.

7. CONCLUSIONS

Wireless technologies play an increasingly critical element in the Chinese urban and infrastructure developments. Local governments and major network operators have increasingly embraced the notion of wireless city, as the access to Internet through public Wi-Fi has become an integral part of a city's economic competitiveness [70,71]. This research traces the recent wireless city development in Shanghai using data gathered from both the government and the telecommunication providers and several interesting findings are uncovered. First, a substantial amount of public Wi-Fi facilities has been deployed and built up, shaping Wi-Fi geographies in Shanghai at multiple spatial scales [69], and a mix of market participants and governmental agencies has been involved in the process of wireless city development in Shanghai. Second, by mapping public Wi-Fi at both district and sub-district levels, we find that both local governments and major telecommunication companies, such as CMCC, have significantly contributed to the widespread Wi-Fi access to the Internet in Shanghai. Results also suggest that telecommunication companies, such as CMCC, have played a more proactive role in the deployment of WLAN hotspots. These findings are similar to the recent wireless development in other cities in Asia, such as Singapore and Taipei [5], while the results also pinpoint the unique characteristics of the Chinese government in the pursuit of wireless city development under rapid urban growth and economic transition. As shown in our micro-scale analysis, Naoqiao New Town, as a major new business community in Shanghai, has fueled the new wireless development by having both governmental agencies and large-size business corporations involved. Songjiang University Town presents a typical case that telecommunication operators, such as CMCC, have played a proactive role in shaping Wi-Fi geographies. In contrast, Pengpu Xingchu has exhibited a sluggish wireless development despite its advantageous location in the central area of Shanghai. Third, from the perspective of social inequalities and sustainability, despite the rapid expansion of both CMCC and "i-Shanghai" WLAN hotspots, the digital divide is still evident especially at the community level. Deployment of WLAN hotspots is obviously biased towards business centers and university campuses as well as their surrounding areas in Shanghai. Public Wi-Fi access was rarely found in poor residential areas in the inner city and remote rural areas of Shanghai. This polarized structure indicates that the problem of digital divide is far from being solved while new digital divide has been created, imposing profound challenges for making Shanghai a sustainable wireless city in the future. Finally, although we have mapped the public Wi-Fi access in Shanghai, studies of digital divide are also promising from a user or customer's perspective, especially when this type of data becomes available. In addition, as China is a country characterized by its huge size and regional differentials, more research is needed to compare the development of wireless cities in different geographical regions or across urban hierarchy in China [72].

ACKNOWLEDGMENTS

The authors would like to acknowledge the funding of the National Natural Science Foundation of China (41371175, 41329001), and the Ford Foundation (0155-0883). This study was also supported by the Key Project of Key Research Institute of Humanities and Social Sciences at Universities (13JJD840010) funded by the Ministry of Education of China. Felix Liao would like to acknowledge partial support from the Idaho EPSCoR (Experimental Program to Stimulate Competitive Research) Program and by the National Science Foundation under award number IIA-1301792.

AUTHOR CONTRIBUTIONS

Mingfeng Wang, Felix Liao, Juan Lin, and Chengcheng Gu collected and analyzed the data. Felix Liao, Mingfeng Wang, and Li Huang wrote the manuscript. Dennis Wei edited and revised the paper. All authors read and approved the manuscript.

REFERENCES

1. Castells, M.; Fernandez-Ardevol, M.; Qiu, J.L.; Sey, A. Mobile Communication and Society: A Global Perspective; Mit Press: Cambridge, MS, USA, 2009.
2. Ishida, T.; Isbister, K. Digital Cities: Technologies, Experiences, and Future Perspectives; Springer Science & Business Media: Berlin, Germany, 2000.
3. Komninos, N. Intelligent Cities and Globalisation of Innovation Networks; Routledge: London, UK, 2008; Volume 37, pp. 1–307.

4. Sheller, M.; Urry, J. Mobile Technologies of the City; Routledge: London, UK, 2006.
5. Hu, M.C. Antecedents for the adoption of new technology in emerging wireless cities: Comparisons between singapore and taipei. Reg. Stud. 2014, 48, 665–679.
6. Shin, D.-H. Ubiquitous city: Urban technologies, urban infrastructure and urban informatics. J. Inf. Sci. 2009, 35, 515–526.
7. Caragliu, A.; del Bo, C.; Nijkamp, P. Smart cities in europe. J. Urban Technol. 2011, 18, 65–82.
8. Gil-Castineira, F.; Costa-Montenegro, E.; Gonzalez-Castano, F.J.; Lopez-Bravo, C.; Ojala, T.; Bose, R. Experiences inside the ubiquitous oulu smart city. Computer 2011, 44, 48–55.
9. Cieslik, A.; Kaniewsk, M. Telecommunications infrastructure and regional economic development: The case of poland. Reg. Stud. 2004, 38, 713–725.
10. Pearson, I. The role of future ict in city development. Foresight J. Futur. Stud. Strateg. Think. Policy 2006, 8, 3–16.
11. Alfirevic, N.; Pavicic, J.; Znidar, K. Knowledge and Information Technology and Urban (Regional) Development; World Scientific and Engineering Acad and Soc: Athens, Greece, 2009; pp. 112–116.
12. Gunasekaran, V.; Harmantzis, F.C. Emerging wireless technologies for developing countries. Technol. Soc. 2007, 29, 23–42.
13. Gaspar, J.; Glaeser, E.L. Information technology and the future of cities. J. Urban Econ. 1998, 43, 136–156.
14. CNN. China's Online Users more than Double Entire U.S. Population. Available online: http://www.cnn.com/2015/02/03/world/china-internet-growth-2014/ (accessed on 2 October 2015).
15. China Internet Network Information Center. Statistical Report on Internet Development in China; CNNIC: Beijing, China, 2014.
16. CNNIC. The 33rd Statistical Report on Internet Development in China; China Internet Network Information Center: Beijing, China, 2014.
17. CNNIC. Development Reports of Shanghai Internet, 2012; China Internet Network Information Center: Shanghai, China, 2012. [Google Scholar]
18. CNNIC. Development Reports of Shanghai Internet, 2009; China Internet Network Information Center: Shanghai, China, 2009.
19. Piro, G.; Cianci, I.; Grieco, L.A.; Boggia, G.; Camarda, P. Information centric services in smart cities. J. Syst. Softw. 2014, 88, 169–188.

20. Steenbruggen, J.; Tranos, E.; Nijkamp, P. Data from mobile phone operators: A tool for smarter cities? Telecommun. Policy 2015, 39, 335–346.
21. Qi, L.; Zeng, J.; Zhang, Y. Study on the construction and operation in wireless city. In Electronic Information and Electrical Engineering; Chen, B.J., Hu, V., Kong, D., Eds.; Atlantis Press: Paris, France, 2012; Volume 19, pp. 70–74.
22. Le, A.N.H.; Wang, M.C.H.; Cheng, J.M.S.; Huang, S.C.F. The mobility value of internet services in a wireless city: The case of taipei, taiwan. Asian J. Technol. Innov. 2012, 20, 51–65.
23. Berry, M.; Hamilton, M. Changing urban spaces: Mobile phones on trains. Mobilities 2010, 5, 111–129.
24. Wilken, R. A community of strangers? Mobile media, art, tactility and urban encounters with the other. Mobilities 2010, 5, 449–468.
25. Rettie, R. Mobile phones as network capital: Facilitating connections. Mobilities 2008, 3, 291–311.
26. Choi, J.H.-J. The city is connections: Seoul as an urban network. Multimedia Syst. 2010, 16, 75–84.
27. Sandvig, C. An initial assessment of cooperative action in Wi-Fi networking. Telecommun. Policy 2004, 28, 579–602.
28. Sedgley, N.; Elmslie, B. Do we still need cities? Evidence on rates of innovation from count data models of metropolitan statistical area patents. Am. J. Econ. Sociol. 2011, 70, 86–108.
29. Helgason, O.; Kouyoumdjieva, S.T.; Karlsson, G. Does mobility matter? In Proceedings of the Wons 2010: Seventh International Conference on Wireless on-Demand Network Systems and Services, Kranjska Gora, Slovenia, 3–5 February 2010; pp. 9–16.
30. Ballon, P.; van Audenhove, L.; Poel, M.; Staelens, T. Business models for wireless city networks in the eu and the us: Public inputs and public leverage. In Telecommunication Markets: Drivers and Impediments; Curwen, P., Preissl, B., Haucap, J., Eds.; Springer: New York, NY, USA, 2009; pp. 325–340.
31. Grubesic, T.; Murray, A. "Where" matters: Location and Wi-Fi access. J. Urban Technol. 2004, 11, 1–28.
32. Oyana, T.J. Exploring geographic disparities in broadband access and use in rural southern illinois: Who's being left behind? Gov. Inf. Q. 2011, 28, 252–261.

33. Driskell, L.; Wang, F. Mapping digital divide in neighborhoods: Wi-Fi access in baton rouge, louisiana. Ann. GIS 2009, 15, 35–46.
34. Ford, G.S.; Koutsky, T.M. Broadband and economic development: A municipal case study from florida. Rev. Urban Reg. Dev. Stud. 2005, 17, 216–229.
35. Torrens, P.M. Wi-Fi geographies. Ann. Assoc. Am. Geogr. 2008, 98, 59–84.
36. Evans-Cowley, J. Planning in the real-time city: The future of mobile technology. J. Plan. Lit. 2010, 25, 136–149.
37. Girth, A.M. What drives the partnership decision? Examining structural factors influencing public-private partnerships for municipal wireless broadband. Int. Public Manag. J. 2014, 17, 344–364.
38. Park, N.; Lee, K.M. Wireless cities: Local governments' involvement in the shaping of Wi-Fi networks. J. Broadcast. Electron. Media 2010, 54, 425–442.
39. Soh, P.-H.; Yu, J. Institutional environment and complementary assets: Business strategy in china's 3g development. Asia Pac. J. Manag. 2010, 27, 647–675.
40. Castells, M. European cities, the informational society, and the global economy. Tijdschr. Voor Econ. En Soc. Geogr. 1993, 84, 247–257.
41. Cartier, C.; Castells, M.; Qiu, J.L.C. The information have-less: Inequality, mobility, and translocal networks in chinese cities. Stud. Comp. Int. Dev. 2005, 40, 9–34.
42. Harvey, D. Social-justice, postmodernism and the city. Int. J. Urban Reg. Res. 1992, 16, 588–601.
43. Grubesic, T.H.; Murray, A.T. Waiting for broadband: Local competition and the spatial distribution of advanced telecommunication services in the united states. Growth Chang. 2004, 35, 139–165.
44. Kellerman, A. Mobile broadband services and the availability of instant access to cyberspace. Environ. Plan. A 2010, 42, 2990–3005.
45. Perrons, D. Understanding social and spatial divisions in the new economy: New media clusters and the digital divide. Econ. Geogr. 2004, 80, 45–61.
46. Castells, M. The Internet Galaxy. Reflections on the Internet Business and Society; Oxford University Press: Oxford, UK, 2001.
47. Fuentes-Bautista, M.; Inagaki, N. Reconfiguring public internet access in austin, tx: Wi-fi's promise and broadband divides. Gov. Inf. Q. 2006, 23, 404–434.

48. Sahiner, T.; Ozbakir, A.; Kurt, G.K. Intra-city digital divide measurements through clustering. In Proceedings of 2013 ITU Kaleidoscope: Building Sustainable Communities (K-2013), Kyoto, Japan, 22–24 April 2013; pp. 233–238.
49. Strover, S.; Mun, S.-H. Wireless broadband, communities, and the shape of things to come. Gov. Inf. Q. 2006, 23, 348–358.
50. Prieger, J.E. The broadband digital divide and the benefits of mobile broadband for minorities. J. Econ. Inequal. 2015, 13, 373–400. Myeong, S.; Kwon, Y.; Seo, H. Sustainable e-governance: The relationship among trust, digital divide, and e-government. Sustainability 2014, 6, 6049.
51. Mingfeng, W.; Chengcheng, G. Building a smart city: A spatial analysis and review of wlan hotspots in metropolitan shanghai. Prog. Geogr. 2015, 34, 438–447.
52. Li, R.; Shiu, A. Internet diffusion in china: A dynamic panel data analysis. Telecommun. Policy 2012, 36, 872–887.
53. Feng, G.C. Factors affecting internet diffusion in china: A multivariate time series analysis. Telemat. Inf. 2015, 32, 681–693. Shanghai CMCC. Available online: www.10086.cn/sh/ (accessed on 18 January 2016).
54. The "i-Shanghai" Project. Available online: http://i-shanghai.sheitc.gov.cn (accessed on 18 January 2016).
55. Wei, Y.D.; Bi, X.; Wang, M.; Ning, Y. Globalization, economic restructuring, and locational trajectories of software firms in shanghai. Prof. Geogr. 2015, 1–16.
56. Liu, G.Z.; Zhou, Q. An empirical analysis on industrial structural transition and regional inequality in china. In Proceedings of the 2009 International Conference on Management Science and Engineering, Moscow, Russia, 14–16 September 2009; Lan, H., Ed.; IEEE: New York, NY, USA, 2009; pp. 1044–1052.
57. Getis, A.; Ord, J.K. The analysis of spatial association by use of distance statistics. Geogr. Anal. 1992, 24, 189–206.
58. Ord, J.K.; Getis, A. Local spatial autocorrelation statistics: Distributional issues and an application. Geogr. Anal. 1995, 27, 286–306.
59. Shen, J.; Wu, F. Moving to the suburbs: Demand-side driving forces of suburban growth in china. Environ. Plan. A 2013, 45, 1823–1844.
60. Walcott, S.M.; Pannell, C.W. Metropolitan spatial dynamics: Shanghai. Habitat Int. 2006, 30, 199–211.
61. He, D.; Ning, Y. Evolution of population structure and spatial distribution in shanghai since 2000. In Urban Development Challenges,

Risks and Resilience in Asian Mega Cities; Springer: Berlin, Germany, 2015; pp. 227–249.
62. Chen, L.; Song, S.; Wang, L. Low-Carbon City and Complex Community: From Planning Ideas to Planning Practice—A Case Study of Shanghai Nanqiao New City. Urban Plan. Forum 2011, 4, 1–8.
63. Wang, R. Planning and design of wireless city based on wlan. Telecommun. Sci. 2011, 6, 21–27. (In Chinese).
64. Li, Z.; Wu, F. Socio-spatial differentiation and residential inequalities in shanghai: A case study of three neighbourhoods. Hous. Stud. 2006, 21, 695–717.
65. Cheng, H.; Wang, M. Analysis on factors affecting floating population's using of internet at home: Insights from a survey of urban villages in shanghai. China Urban Stud. 2012, 5, 81–93. (In Chinese).
66. Wei, Y.D. Planning chinese cities: The limits of transitional institutions. Urban Geogr. 2005, 26, 200–221.
67. Wei, Y.D. Zone fever, project fever: Development policy, economic transition, and urban expansion in china. Geogr. Rev. 2015, 105, 156–177.
68. Macagnano, E.V. Intelligent urban environments: Towards e-inclusion of the disabled and the aged in the design of a sustainable city of the future.
69. A south african example. In Sustainable City V: Urban Regeneration and Sustainability; Gospodini, A., Brebbia, C.A., Eds.; Wit Press/Computational Mechanics Publications: Southampton, UK, 2009; Volume 117, pp. 537–547.
70. Roper, S.; Grimes, S. Wireless valley, silicon wadi and digital island - helsinki, tel aviv and dublin and the ict global production network. Geoforum 2005, 36, 297–313
71. Mack, E.A. Variations in the broadband-business connection across the urban hierarchy. Growth Chang. 2015, 46, 400–423.

8

CHAPTER

AN INTERACTIVE WIRELESS COMMUNICATION SYSTEM FOR VISUALLY IMPAIRED PEOPLE USING CITY BUS TRANSPORT

Hsiao-Lan Wang [1], Ya-Ping Chen [2], Chi-Lun Rau [2,†,*] and Chung-Huang Yu [1,†]

ABSTRACT

Visually impaired people have difficulty accessing information about public transportation systems. Several systems have been developed for assisting visually impaired and blind people to use the city bus. Most systems provide only one-way communication and require high-cost and complex equipment. The purpose of this study is to reduce the difficulties faced by visually impaired people when taking city buses, using an interactive wireless communication system. The system comprised a user module and a bus module to establish a direct one-to-one connection. When the user inputs 4-digit numbers, the user module immediately sends out the information. If the bus module receives the matched bus number, it buzzes and the warning LED flashes to notify the bus driver that someone is waiting to board on the bus. User tests were conducted by two visually impaired people in a simulated vehicle and a city bus. The success rate of interactive wireless communication, recognizing the arrival of the bus and boarding the correct bus reached 100% in all of the tests. The interactive wireless communication aid system is a valid and low-cost device for assisting visually impaired people to use city buses.

Keywords: wireless communication; visual impairment; public transportation

1. INTRODUCTION

The use of public transport is vital to the productivity and independence of visually impaired people. Helping visually impaired people use public transport can increase their chances of education and employment and reduce the financial burden on their families [1,2]. In most physical environments, the visually impaired have difficulty accessing information about transport stops, terminals, vehicles, schedules, maps, and directories, which prevent them from using public transport effectively. According to a survey in Taiwan on the living demands of disabled people, using public transport was the most critical problem for the visually impaired, amounting to 71.04% of 602 visually impaired people [3]. The survey results showed that only 14% of visually impaired people used public transport (city bus, mass rapid transit, train, *etc.*).

Knowing the location of the bus stop and the time when the bus arrives are two common difficulties faced by the visually impaired. Some position navigation systems have been developed to solve the problem of locating bus stops [4,5,6]. Advanced public transportation services (APTS), including bus dynamic information display systems with the Global Positioning System (GPS) technology, have been developed by many countries. APTS can be equipped with the bus-stop voice reporting systems to provide more information to visually impaired people on the arrival of the bus they want to board. Some APTS combining special handheld devices were designed to provide the electronic orientation and dynamic information for visually impaired people. Examples include Apex in Czech [7], Bus-ID in German [8], iBus in Canada [9], NOPPA in Finland [10], PAVIP (Personal Assistant for Visual Impaired People) in Switzerland [11], and Quo Vadis in Austria [12]. Recently, smart phones become more and more widespread. Markiewicz and Skomorowski proposed an aid system using mobile phones (Java and GPS-function enabled) as the user modules [13]. Some smart phone apps providing local bus information through special or voice interface are developed for the visual impaired like Geogiephone [14], PT Guide [15], and Voice@Bus [16].

In Taiwan, the Taipei city government once utilized an experimental system named "broadcasting bus", announcing the bus number upon arriving at the bus terminal, to serve the visually impaired in 2002 [17]. Now there are about 300 bus routes, 4,000 buses, and 5,000 bus stops in Taipei city [18,19]. The APTS named "e-bus system" started to be deployed in Taipei city since 2005. The total budget was about 7 million U.S. dollars [20]. The GPS position of the buses was transmitted to a server at a control center by general packet radio service (GPRS) technology, and the related information was transmitted to the intelligent bus terminals by GPRS technology immediately. The number and the waiting time of the buses were shown on the light-emitting diode (LED) screen at the smart bus stops. Totally 1000 smart bus stops, 20 percent of all bus stops, is planned to build. A pilot survey of 400 passengers in 1998 showed that the smart bus stops are not adequate when equipped only with the voice reporting system because voice information was confusing when many buses approached the same terminal at the same time [21].

The APTS with bus-stop voice reporting system and/or handheld devices might not provide a comprehensive solution to this important issue. A survey on visually impaired people showed that it would be very helpful if somebody could alert bus drivers about their boarding on buses [22]. Various systems have been developed for visually impaired and blind people to communicate with bus drivers. Mehra *et al.* developed a user-triggered bus identification system in 2010 [23]. The user could select a particular bus and send signals by user module, and then a small bulb starts flickering in the driver's control panel. This system offered only one-way communication. Bischof *et al.* developed a wireless local area network (WLAN) communication system named NAVCOM [24]. The authors proposed that blind people need a feedback to acknowledge whether the bus drivers get the original message. Another bus identification system designed by El Alamy *et al.* involves a bus station controller to recognize users and send signals to buses with radio frequency [25]. This system will announce the information of the bus number when there is a 2-meter distance between the bus and the bus station. In Taipei, bus drivers are inclined to leave the bus stop as soon as possible because most time several buses simultaneously arrive at the same bus stop. Visually impaired people are often ignored at bus stops if no one informs bus drivers about waiting passengers behind. The purpose of the present study is to reduce the difficulties faced by visually impaired people when taking buses with interactive wireless communication design. The interactive feedback mechanism would allow visually impaired people and bus drivers to receive the transmitted signals from each other and improve the success rate of boarding correct buses.

2. MATERIALS AND METHODS

2.1. Participants

The participants in this study were two visually impaired people, participant A was one 52 year-old male and participant B was one 50 year-old female. The male subject had been totally blind for 20 years as a result of retinitis pigmentosa, without any other disability. The female subject's total blindness was congenital without any other disability. Both participants were independently mobile with the use of a guide cane.

2.2. System Development

In a real environment, the interaction between visually impaired people and bus drivers is a many-to-many relationship rather than a one-to-one. For technological simplicity, this system used a one-to-one interactive communication system. The collision of communication data over the one-to-one interactive wireless transmission leads to the poor stability and low accuracy of wireless communication. Wireless communication technology that do not experience collision include time division multiple access (TDMA), code division multiple access (CDMA), and frequency-division multiple access (FDMA). This study primarily refers to the principles and concepts of FDMA, that is, the signal communicating through two distinct frequency bands: 434 MHz and 315 MHz.

The interactive wireless communication system would have two modules: a user module for the visually impaired and a bus module for bus drivers (Figure 1). The hardware was developed according to the interactive design. The user module (Figure 2) consists of a single-chip CPU (PSoC®CY8C27443), a wireless transmission module (434 MHz), a wireless receiving module (315 MHz), a coding IC (HT-12E), a decoding IC (HT-12D), a buzzer, a keypad, an LCD, and a power LED. The bus module (Figure 3) consists of a single-chip CPU (PSoC®CY8C27443), a wireless transmission module (315 MHz), a wireless receiving module (434 MHz), a coding IC (HT-12E), a decoding IC (HT-12D), a buzzer, an LCD, a warning LED, and a power LED. When the user module sends out a signal, it is transmitted through the 434 MHz band to the bus module, which then replies through the 315 MHz band to the user module. In this way, the problem of wireless data collisions between users is avoided.

A prototype of user module was tested by both participants. After testing the prototype, the participants suggested several areas of improvement, including the device's size, operational inconvenience—particularly the power switch and button settings, and low buzz volume. The devices were improved according to the users' feedback. The total price of our system is about 120 U.S. dollars.

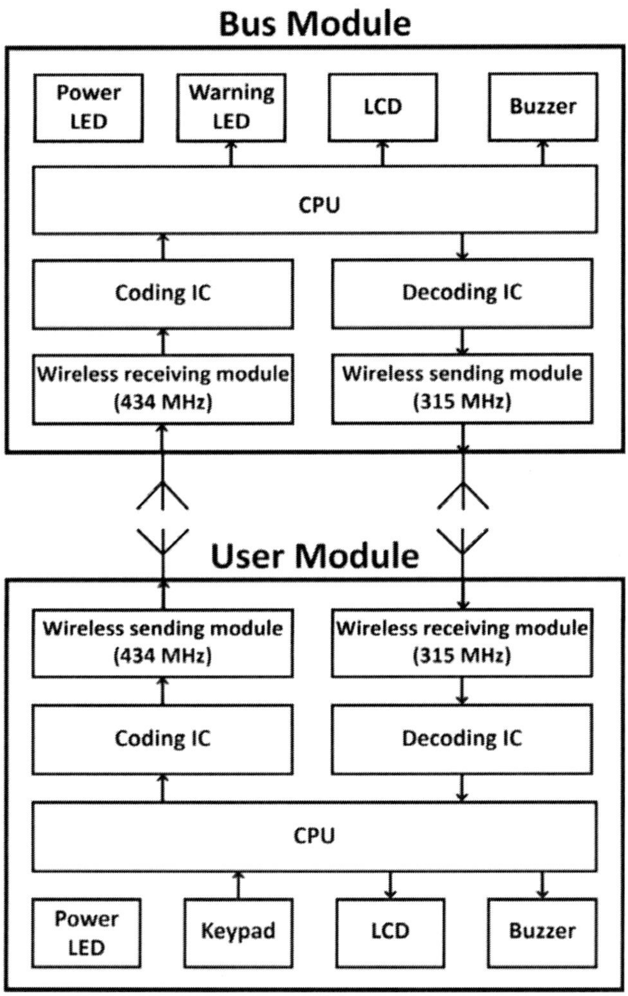

Figure 1. System framework of the user module and the bus module.

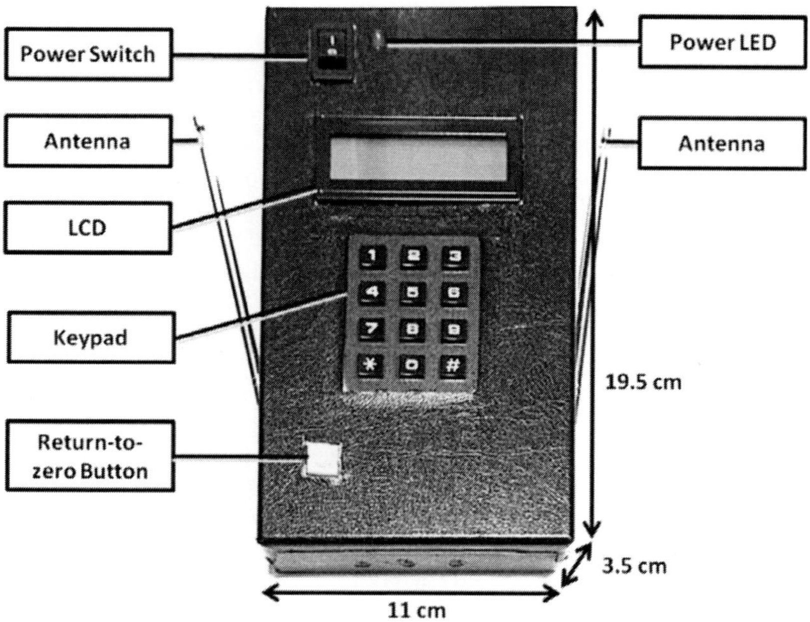

Figure 2. User module.

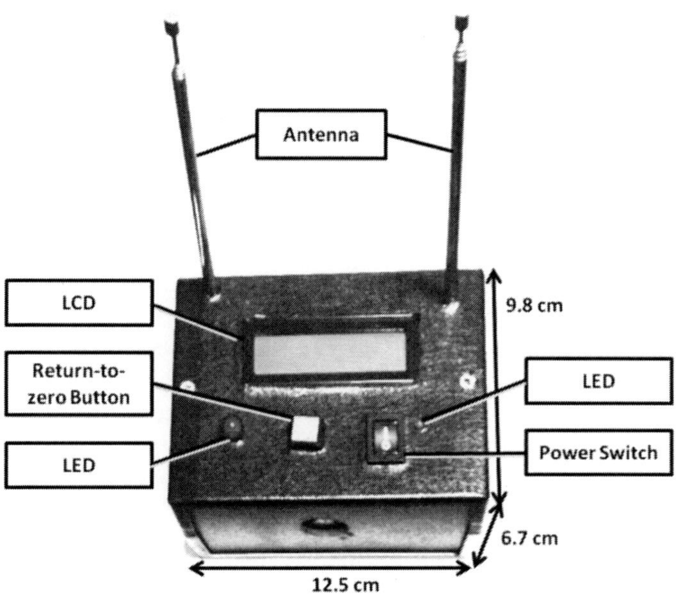

Figure 3. Bus module.

2.3. Module Interface and Interactive Wireless Communication Design

When the user module power is switched on, the power LED is on, and the LCD displays "Bus No?" to prompt the user to input the preset bus numbers. The module can store 16 sets of 4-digit numbers, for example: #074, *074, and *001. The sign # and * represent the traveling direction of city buses, and the number behind the sign represents the bus number. The indication of the direction is necessary because buses with the same number traveling in different directions will receive the signal at the same time. When the user has successfully input one of the preset 4-digit numbers, the module immediately sends out the information through the wireless transmission module continuously. Because this system was designed to assist visually impaired people, the keypad has audible tones. Additionally, if the input number was not the same as the preset number, the buzzer of the user module emits a specific sound to inform the user that the number is incorrect. However, if the input number is correct, no sound is emitted. When the target bus is coming, and the signal of the correct bus number is received, the buzzer is activated and the bus number is displayed on the LCD to inform the user that the bus is arriving.

16 sets of 4-digit numbers representing bus traveling directions and bus numbers are also preset in each bus module. When the power of the bus module is switched on, the power LED is on, and the LCD displays "Bus Number: #xxx" or "Bus Number: *xxx". The sign # and * represent the traveling direction of city buses, and the xxx behind the sign represents the bus number. If the bus module receives the correct bus number, the bus module buzzes and the warning LED flashes, and the LCD shows "Attention!" to notify the bus driver that someone is waiting to board that particular bus. The bus module also sends the same bus number back to the user module at the same time. When the bus drivers have taken on the passenger, they can press the return-to-zero button to turn off the warning LED and the buzzer. Then the bus module is available for the next passenger. The user can press the return-to-zero button to turn off the warning LED and the buzzer after the arrival of the bus. Then the user module is available for inputting the next set of numbers.

2.4. Outcome Measure

A basic requirement of this system was the wireless communication distance. If the wireless communication distances were shorter than buses' stopping distances, bus drivers would fail to stop buses for visually impaired people. The stopping distance of a car is equal to the reaction time plus the braking distance. Stopping distance can be affected by vehicle weight, weather, road condition, and speed. The speed limit for city buses in Taipei is 40 km/h, but most buses always travel at a speed less than 40 km/h and slow down when approaching bus stops. Because the users were waiting at bus stops, we decided to keep the simulated vehicle in tests at 30 km/h. The stopping distance of a bus at 30 km/h on dry roads is approximately 12.1 meters [26], and this was the parameter for our tests. Conversely, if the wireless communication distance was too long, it was possible that the bus driver would have received the information from a waiting user too early. To avoid such situations, the maximum transmission distance of this system must be less than 100 meters.

The following tests measured the wireless communication distance, the success rate of the interactive wireless communication, and visually impaired people recognizing arrival of the bus and boarding the correct bus.

2.5. Test Procedure

Test 1 was conducted on a city road with few vehicles, and measured the accuracy of the interactive wireless communication under various weather conditions. One experimenter held the user module and input the bus numbers, while the other experimenter took the bus module and drove a motorcycle at 30 km/h to the former from a long distance away. When the bus module buzzed and the warning LED flashed, the distance was recorded. This test was repeated five times in sunny, cloudy, and rainy weather.

Test 2 was conducted on a campus road, with few vehicles present, and in sunny weather. Before the tests, it took the two participants approximately three to five minutes to learn how to use the user module. The participants held the user module at a fixed location. The experimenter drove a sports utility vehicle (SUV), equipped with the bus module, as a simulated bus. The starting distance between these two modules was 150 m. The driver of the SUV approached at 30 km/h, while the participant input the bus number. When the bus module buzzed and the warning LED flashed, the distance between the bus module and the user module was recorded. This test was repeated five times for each participant.

Test 3 took place on a city road in sunny weather. Before the tests, it took the two participants approximately three to five minutes to learn how to use the user module. The participants held the user module at a bus stop. The experimenter was on board a bus equipped with the bus module, a few stops before the participants. The driver of the city bus was driving normally, while the participant was inputting the bus number. When the bus module buzzed and the warning LED flashed, the distance between the bus module and the user module was recorded by the experimenter. These tests were repeated five times for each participant.

We defined the success of interactive wireless communication as the user module buzzed in response to the bus module.

3. RESULTS

The results of the tests are shown in Table 1. In all three tests, the success rates of the interactive wireless communication, recognition of the bus' arrival, and boarding of the correct bus were all 100%. The maximum wireless communication distances of these tests were below 100 meters. In Test 1, the average, maximum, and minimum wireless communication distance was the longest when the weather was sunny, followed by cloudy days and then by rainy days. The minimum communication distance was greater than 12.1 meters in sunny and cloudy weather, but failed in rainy weather. In Test 2, the minimum wireless communication distance was longer than 12.1 meters for both participants. In Test 3, the minimum wireless communication distance for both participants was shorter than 12.1 meters, but the average distance was longer than 12.1 meters for participant B. The bus traveled at a speed of approximately 20 to 30 km/h.

4. DISCUSSION

Our low-cost system facilitated successful communication and feedback between visually impaired people and bus drivers. Using the interactive communication modules, visually impaired people can give information to bus drivers actively and board the correct bus easily. Bus drivers can receive notification early and send feedback to passengers automatically. This system created direct and fast wireless communication between visually impaired people and city buses drivers.

The interactive wireless communication was successful in all tests, meaning the concept of FDMA was suitable for this system. However, in Test 1, wireless communication was substantially influenced by atmospheric attenuation. We also observed that when the antenna of the user module is perpendicular to the ground, the wireless communication distance was the

Table 1. Results of Test 1 to Test 3.

Test	One			Two		Three	
Participant	Experimenter	Experimenter	Experimenter	Visual Impaired A	Visual Impaired B	Visual Impaired A	Visual Impaired B
Weather	Sunny	Cloudy	Rainy	Sunny	Sunny	Sunny	Sunny
Vehicle	Motorcycle	Motorcycle	Motorcycle	SUV	SUV	Bus	Bus
Speed (km/h)	30	30	30	30	30	20–30	20–30
Success rate (%)							
Wireless communication	100	100	100	100	100	100	100
Recognizing arrival of the bus	-	-	-	100	100	100	100
Boarding the correct bus	-	-	-	100	100	100	100
Wireless communication distance (m)							
Minimum	35.0	15.0	8.0	15.0	25.0	5.0	10.0
Maximum	45.0	35.0	15.0	50.0	33.0	15.0	25.0
Average (mean±SD)	40.0 ± 5.0	26.7 ± 7.4	11.0 ± 2.7	29.4 ± 13.5	29.6 ± 3.2	10.0 ± 5.0	16.7 ± 7.6

Notes: SUV, sports utility vehicle. SD, standard deviation.

longest. The position of the user module changed the direction of the antenna and influenced the wireless communication distance. This might be the reason for differences in the minimum and maximum wireless communication between participants in Tests 2 and 3. Further improvement of the wireless communication is necessary before commercialization of this system is possible.

In Test 3, the minimum wireless communication distance for both participants was less than 12.1 meters. Compared with the results of Test 2, the wireless communication distance in Test 3 was an average of 54.8% shorter. This result might be attributed to the bus windows being closed and to the city roads being surrounded by large buildings. Although the wireless communication distance was insufficient, both participants were able to successfully board the correct bus. We observed the bus travelling at a speed of approximately 20 km/h when approaching the stop. The stopping distance of a bus at 20 km/h on dry roads is approximately 6.0 meters [26]. The driver was able to stop the bus quickly and easily.

In Taipei, bus drivers are inclined to stop the bus at bus terminals only once passengers have indicated that they want to board. Therefore, visually impaired people need to hold a large piece of paper with the bus number written on it at bus terminal, or ask other passengers to inform them the arrival of the bus. We interviewed the bus drivers about the experience of using the bus module after Test 3. The bus drivers believed this interactive communication aid system could alert drivers to visually impaired people who are waiting for the arrival of a bus and reduce the drivers' mental workloads.

After Test 3, the participants provided their suggestions for the interactive wireless communication system. Both of the visually impaired participants believed the system would help to differentiate between several buses arriving simultaneously. They agreed that the bus or bus-stop voice reporting systems are not suitable for a noisy environment in Taipei. If the visually impaired can give information to bus drivers actively, they would be less likely to be ignored at bus stop. The participants also wished the market price of the user module is less than 100 U.S dollars. Regarding the interface of the user module, they agreed that it was an easily operated system. Regarding the hardware design, they suggested a smaller user module with voice keypad and voice feedback. They also mentioned about the combination of the user module and the smart bus stops. If succeeded, they don't have to bring the user module all the time.

Further improvements that need to be made to the system include: (1) improving the stability of the wireless communication to reduce factors, such as weather and the neighboring environment, affecting the interactive wireless communication distance; (2) upgrading the system to a multi-pair and multi-direction wireless communication system that meets the requirements of a larger group of visually impaired people taking multiple buses at a given time; (3) upgrading the sound feedback of the keypad on the user module to an automated voice keypad and providing voice instructions. Improving these functions will allow for a more fully developed interactive wireless communication system.

5. CONCLUSIONS

An interactive wireless communication aid system for the visually impaired to use city buses was developed in this study. Results of tests indicated that this system could help users to successfully board their desired buses, using the interactive communication modules, which worked most favorably in sunny weather. The interface design of this system is simple to use and easy to understand. The research demonstrated the feasibility of the proposed system, and provided a reference basis for developing a new system for aiding visually impaired bus users in the future.

ACKNOWLEDGMENTS

The authors thank the participants.

AUTHOR CONTRIBUTIONS

Hsiao-Lan Wang designed and carried out the study and drafted the paper. Ya-Ping Chen assisted the data analysis and managed the manuscript. Chi-Lun Rau assisted the first author building the research framework and carried out the study. Chung-Huang Yu developed the hardware system and took charge of the whole framework of this work.

REFERENCES

1. Montarzino, A.; Robertson, B.; Aspinall, P.; Ambrecht, A.; Findlay, C.; Hine, J.; Dhillon, B. The impact of mobility and public transport on the independence of visually impaired people. *Vis. Impair. Res.* **2007**, *9*, 67–82.
2. Neuville, E.; Izaute, M.; Trassoudaine, L. A wayfinding pilot study: The use of the Intelligent Public Vehicle by people with visual impairment. *Br. J. Vis. Impair.* **2009**, *27*, 65–74.
3. Ministry of the Interior. Survey on Difficulties in Dailiy Living of above 6 Year-Old Physically and Mentally Disabled. In *Report on Physically and Mentally Disabled Citizens Living Demand Survey*; Ministry of the Interior Taiwan: Taipei, Taiwan, 2006; p. 329.
4. Wiener, W.R.; Ponchillia, P.; Joffee, E.; Rutberg-Kuskin, J.; Brown, J. The effectiveness of external bus speaker systems for persons who are visually impaired. *J. Vis. Impair. Blind.* **2000**, *94*, 421–433.
5. Provost-Hatlen, T.; Myers, L.A. Advocating in Bahalf of Blind and Visually Impaired Bus Travelers. In *Access to Mass Transit for Blind and Visually Impaired Travelers*; Uslan, M.M., Peck, A.F., Wiener, W.R., Stern, A., Eds.; American Foundation for the Blind: New York, NY, USA, 1990; pp. 87–91.
6. Passenger BUS Alert System for Easy Navigation of Blind. Available online: http://www.kresttechnology.com/krest-academic-projects/ krest-mtech-projects/ECE/M.TECH%20EMBEDDED-ARM7/BASEPAPERS/31.PASSENGER%20BUS%20ALERT%20SYSTEM %20FOR%20EASY%20NAVIGATION%20OF%20BLIND.pdf (accessed on 20 February 2014).
7. The Electronic Orientation and Information System For The Visually Impaired Persons. Available online: http://www.apex-jesenice.cz /index.php?lang=en (accessed on 31 March 2014).
8. BUS-ID: RFID-Basierte Akustische Mobilitatsunterstutzung. Available online: http://www.hsu-hh.de/aut/index_Cpb5mbolsp7SBzgx.html (accessed on 31 March 2014).
9. iBUS. Available online: http://www.stm.info/en/about/ major_projects/ibus (accessed on 31 March 2014).

10. NOPPA—Navigation and Guidance System for the Blind in Finland. Available online: http://www.eltis.org/index.php?id= 13&study_id= 1751 (accessed on 31 March 2014).
11. PAVIP Transport. Available online: http://www.bones.ch/bones/transport.php (accessed on 31 March 2014).
12. Accessibility in Vienna. Available online: http://www.urlaubfueralle.at/upload/tagung/ENAT09/shared/33.Session-3-Krpata.pdf (accessed on 31 March 2014).
13. Markiewicz, M.; Skomorowski, M. Public transport information system for visually impaired and blind people. *Transp. Syst. Telematics* **2010**, *104*, 271–277.
14. GeorgiePhone. Available online: http://www.georgiephone.com/ (accessed on 31 March 2014).
15. PT Guide. Available online: http://itunes.apple.com/us/app/from_topapp.net/id455704043?ign-mpt=uo%3D4 (accessed on 31 March 2014).
16. Voice@Bus. Available online: http://play.google.com/store/apps/details?id=rp.fyp.busvoice (accessed on 31 March 2014).
17. Voice Reporting Plan Outside the Bus upon Arriving at the Stop. Available online: http://www.taipei.gov.tw/cgi-bin/Message/ (accessed on 4 May 2009).
18. More Than 500 Smart Bus Stops. Available online: http://www.iot.gov.tw/ct.asp?xItem=528560&ctNode=2016 (accessed on 31 March 2014).
19. Taipei City Bus Information System. Available online: http://www.asiatek.com.tw/casestudies/casestudies01.html (accessed on 31 March 2014).
20. E-Bus System. Available online: http://e-bus.tpc.gov.tw/html/ebus_2.htm (accessed on 8 July 2009).
21. Chang, T.H.; Chiang, I.C.; Yeh, Y.H.; Wu, M.Y.; Kuo, C.T.; Hsu, C.C.; Hsu, Y.T.; Wu, Y.R.; Lin, C.S.; Tsai, Y.S. Development and Demonstration Project of Advanced Bus. Information System and In-Vehicle Safety Equipment (I). Ministry of Transportation and Communication: Taipei, Taiwan, 2003; pp. 131–134.
22. Golledge, R.G.; Marston, J.R.; Costanzo, C.M. Attitudes of visually impaired persons toward the use of public transportation. *J. Vis. Impair. Blind.* **1997**, *91*, 446–459.

23. Mehra, D.; Singh, V.; Paul, R.; Garg, A.; Agarwal, C.; Sharma, V.D.; Dhakar, S.; Arora, P.; Balakrishnan, M.; Paul, K.; *et al.* User Triggered Bus Identification and Homing System: Making Public Transport Accessible for the Visually Challenged. In Proceeding of the 12th International Conference on Mobility and Transport for Elderly and Disabled Persons (TRANSED), Hong Kong, China, 2–4 June 2010.
24. Bischof, W.; Krajnc, E.; Dornhofer, M.; Ulm, M. NAVCOM—WLAN Communication between Public Transport Vehicles and Smart Phones to Support Visually Impaired and Blind People. In *Computers Helping People with Special Needs*; Miesenberger, K., Karshmer, A., Penaz, P., Zagler, W., Eds.; Springer: New York, NY, USA, 2012; Volume 7383, pp. 91–98.
25. El Alamy, L.; Lhaddad, S.; Maalal, S.; Taybi, Y.; Salih-Alj, Y. Bus Identification System for Visually Impaired Person. In Proceeding of Next Generation Mobile Applications, Services and Technologies (NGMAST), Paris, France, 12–14 September 2012.
26. Stopping Distances for Cars and Lorries. Available online: http://www.ukspeedtraps.co.uk/stopping.htm (accessed on 17 June 2013).

9

CHAPTER

WING/WORLD: AN OPEN EXPERIMENTAL TOOLKIT FOR THE DESIGN AND DEPLOYMENT OF IEEE 802.11-BASED WIRELESS MESH NETWORKS TESTBEDS

ABSTRACT

Wireless Mesh Networks represent an interesting instance of light-infrastructure wireless networks. Due to their flexibility and resiliency to network failures, wireless mesh networks are particularly suitable for incremental and rapid deployments of wireless access networks in both metropolitan and rural areas. This paper illustrates the design and development of an open toolkit aimed at supporting the design of different solutions for wireless mesh networking by enabling real evaluation, validation, and demonstration. The resulting testbed is based on off-the-shelf hardware components and open-source software and is focused on IEEE 802.11 commodity devices. The software toolkit is based on an "open" philosophy and aims at providing the scientific community with a tool for effective and reproducible performance analysis of WMNs. The paper describes the architecture of the toolkit, and its core functionalities, as well as its potential evolutions.

1. INTRODUCTION

Wireless Mesh Networks (WMN) [1, 2] represent a technological bridge between mobile ad hoc networks (MANETs) and traditional infrastructure networks, such as the ones based on the IEEE 802.11 family of standards. Compared to infrastructure networks, WMN offer several advantages: (i) they allow the combination of different wireless technologies, such as cellular, WiFi, and WiMAX; (ii) they can be incrementally deployed, in order to gradually extend connectivity and capacity, avoiding massive investments. Unlike the MANET scenario, where all nodes act as both host and routers, in a WMN a distinction exists in terms of functionalities between traffic source/termination points and pure relay devices.

A typical WMN consists of several nodes (routers and gateways) which exploit multihopping in order to build and maintain a wireless backhaul. WMNs enhance traditional star-shaped network architectures by providing increased robustness (e.g., no single points of failure are present and broken/congested links are encompassed), scalability and flexibility (without the need for deploying cables, connectivity may be provided only where and when needed/economically attractive), and incremental deployment. Moreover, WMNs can support heterogeneous transmission technologies.

Mesh routers are typically characterized by a small physical footprint which makes them suitable for a wide range of deployments. As for example, due to their low-energy requirements, mesh routers can be deployed as completely autonomous units with solar, wind, or hydro power. Moreover, WMNs are the perfect enabling technology for community networks, in that their distributed nature lends itself to a decentralized ownership model where each participant owns and maintains his/hers own hardware. Finally, WMNs are also expected to lower the entrance barrier for network operators by allowing them to deploy a wireless backhaul in an incremental fashion.

Albeit, several commercial WMNs are already available [3–7], even at (relatively) low prices, no specific study on the design principles to be followed in order to fully exploit the features of the wireless mesh networking paradigm has been published yet. As a matter of fact, most available works on performance evaluation of WMNs are based on simulation studies, or, in some cases, on analytical frameworks. However, given the complexity and the heterogeneity of the wireless mesh networking scenario and given the high number of functionalities involved—crossing several layers of the protocol stack—it emerges a clear need for real-world deployments and prototypes where novel methodologies and algorithms can be tested and evaluated in an isolated environment.

In this work, an "open" approach is used to design and deploy a wireless mesh networking toolkit where off-the-shelf and low-cost hardware components are used as building core components. As a case study, we discuss the WING/WORLD testbed developed and deployed using the aforementioned toolkit. All the developed software is released under a BSD License [8, 9] and is made freely available to the research community to expand and modify it beyond its current functionalities. The reference technology is IEEE 802.11, but the toolkit can be extended to easily incorporate other wireless (e.g., WiMAX) and wired technologies.

The rest of the paper is organized as follows. Section 2 discusses the most relevant trade-offs that we faced in designing the WING/WORLD toolkit. Section 3 describes the system's concept and architecture. Additional technical details on the features currently supported by the toolkit are given in Section 4, while a comparison with other academic testbeds and prototypes is provided in Section 5. Use cases are presented and discussed in Section 6. Finally, Section 7 draws the conclusions pointing out current and future research directions.

2. PLATFORM DESIGN CHOICES AND TRADE-OFFS

In order to deliver a viable technological solution for ubiquitous wireless network access, WMNs are required to support a broad range of benchmarks and services. Given such a background, testbeds represent the ideal play-ground where innovative solutions can be analyzed in a controlled and realistic environment. As a result, the design of a wireless mesh networking toolkit must be driven by both current research trends and the requirements imposed by the target deployment scenario. This section aims at discussing the most relevant trade-offs that we faced in designing the WING/WORLD platform. We invite the reader to consider the discussed issues not as "closed" topics but instead as starting ground for further investigations.

2.1. Hardware Platforms

In designing a WMN, several issues, ranging from platform selection and node deployment to the selection of a suitable software framework for an efficient and useful testbed operation, must be carefully considered by the network engineer. It is worth noting that none of the aforementioned choices should be considered as the only driving factor in the testbed development. Instead, as we will see in the following sections, they all must be addressed as a coherent solution to a multitude of problems. The choices to be made during the platform selection phase heavily depend on research directions and reference application scenarios in terms of network type and size, expected users, and budget.

Being characterized by costs between 80–100€ (at the time of this writing), home wireless routers (e.g., the Linksys WRT54G) deliver the cheapest solution for wireless mesh networking. The major drawbacks of these devices are the modest processing power, due to CPUs designed for lightweight loads, and the limited radio capabilities, due to small antennas and low power WiFi cards. On the other hand, embedded platforms provide high flexibility in terms of custom and off-the-shelf hardware components and are characterized by a wide performance range. Moreover, outdoor deployment is made easier by tailored water-proof enclosure, Power-Over-Ethernet support, and the absence of any moving part. Embedded platforms based on the x86 architecture (e.g., PCEngines, Soekris, etc.) do not require cross compilation and standard development tools and OSes can be used, while platforms based on non-x86 CPUs (e.g., Gateworks, Routerboard, etc.) provide a better price/performance ratio with the drawback of requiring cross compilation.

It is worth noting that processing power and storage space are hard constraints to both the services and the experiments that can be supported by the network. As a matter of fact, multiradio configurations can easily exceed the CPU capabilities of many embedded platforms especially if traffic forwarding is coupled with synthetic traffic generation. Moreover, a suitable storage space must be provided in order to collect measured data.

2.2. Software Platforms

A wide range of OSes is available for the aforementioned hardware platform, ranging from open-source systems like all Linux variants and *BSD to commercial real-time solutions. Since no currently available software platform may be considered as the final solution for wireless mesh networking, an open-source OS, which makes available the source code under terms that allow modification and redistribution, is therefore the optimal choice to speed up research and prototyping. Linux-based OSes are available for the most important embedded platforms. However, it is worth noting that, albeit these OSes may be very similar to the Linux distributions available on common PCs, the available software packages and the userspace tools and utilities may differ significantly due to both CPU-specific requirements and system resources constraints.

2.3. Routing Frameworks

There are three primary components that influence the routing framework: (i) the protocol architecture, (ii) the routing scheme, and (iii) the software implementation. Routing can be either provided at level three of the ISO/OSI networking stack as modification of the standard IP protocol or by adding an interposition layer between the Data Link Layer and the Network layer. In the latter solution (usually referred to as Layer 2.5 routing), the multihop backhaul is transparent to the upper networking stack, making the WMN appears just as another Ethernet link. On the other hand, such an approach introduces additional encapsulation and processing overhead as a result of, respectively, the header and checksum required by the interposition level. This implies a slight degradation in overall performance, in terms of both throughput and latency. Due to the large availability of routing protocols originating from the MANETs research, Layer 3 routing has been the most popular solution in the earlier WMN implementations. Solutions based on Layer 2.5 routing started mushrooming soon after in both academic and private research institutes (MIT Roofnet and Microsoft MCL are the two most notable precursors). Nowadays, we are observing an increasing trend from commercial vendors toward Layer 2 routing. Such an approach is however typically based on proprietary hardware/software solutions. The authors advocate for Layer 2.5 due to both its enhanced scalability and network stack transparentness.

WMNs share a number of features with ad hoc networks [10]. In particular, WMNs are characterized by self-organization and self-healing capabilities and exploit multihopping to build a wireless backhaul for delivering Internet connectivity to end-users. As a result, many routing protocols developed for Mobile Ad hoc Networks (MANETs) have been adapted to fit the mesh environments. However, as opposed to the MANETs paradigm, research efforts in the WMN community focused on network scalability rather than mobility. As a result, particular attention has been devoted by the academic community to the introduction of novel routing metrics capable of taking into account wireless channel characteristics [11] and to multiradio/multichannel architecture capable of increasing the overall network capacity [12].

Parallels activities aim at addressing the WMNs scalability issues by employing frequency agile/cognitive radios, dynamic spectrum access, and clustering algorithms. In [13] the authors propose the COMNET framework.

COMNET exploits intelligent frequency-shifting self-managed mesh network in order to implement dynamic spectrum sensing and management techniques allowing radios to use frequencies other then those located in the ISM bands. Additionally, in [14] a novel cluster-based middle-ware is proposed. The proposed solution significantly reduces the bandwidth use within the wireless mesh backbone by introducing a clustering service and an adapter. The former builds and maintains clusters of nodes; the latter acts as interface with the applications. However, all these advanced wireless radio technologies and architectures require a revolutionary approach to the communication protocols' design in order to reduce congestion, eliminate potential bottlenecks links, and eventually facilitate the commercialization of WMNs.

Routing protocols are typically generally classified as *proactive*, *reactive*, and *hybrid*. Proactive protocols maintain a list of all destinations and routes while reactive protocols discover routes on-demand when a packet needs to be forwarded. Such a behavior makes proactive routing less suitable for WMNs or in general for networks characterized by low-churn rates (number of nodes that leave the network during a specified time period divided by the average total number of nodes over that same time period). It is the authors' standpoint that on-demand route discovery can result in much less traffic than the standard proactive approach, especially when innovative route caching schemes are employed.

Several proactive and reactive routing protocols are already available for deployment over a WMN. Their implementations differ by code maturity, license, and degree of modification to the standard networking stack. In such a context, moving the routing logic from kernel-space into user-space libraries offers considerable advantages in terms of both faster development cycle and easier debugging at the expense of performance reduction. Due to such considerations, many academic research prototypes exploit routing protocol implementations running in the user-space.

3. CONCEPT AND SYSTEM ARCHITECTURE

The WING/WORLD testbed is an experimental IEEE 802.11 wireless mesh network built using off-the-shelf components. Specific attention is aimed at providing a solution that researchers around the world can easily replicate at their premises and possibly connect to the existing infrastructure to enable to enlarge the test-site. It is worth mentioning that all the developed software has been released under a BSD License and is made fully available to the research community [8, 9].

Current configuration is based on 23 nodes deployed across two buildings, implementing two local indoor wireless mesh networks interconnected by an outdoor WiFi point-to-multipoint wireless link. The geographical distribution of the testbed is presented in Figure 1. The system design was driven by our previous work on the state-of-the-art solutions for engineering a WMN testbed [15] by applying the observed guidelines to a real-world scenario, namely, the WING/WORLD testbed.

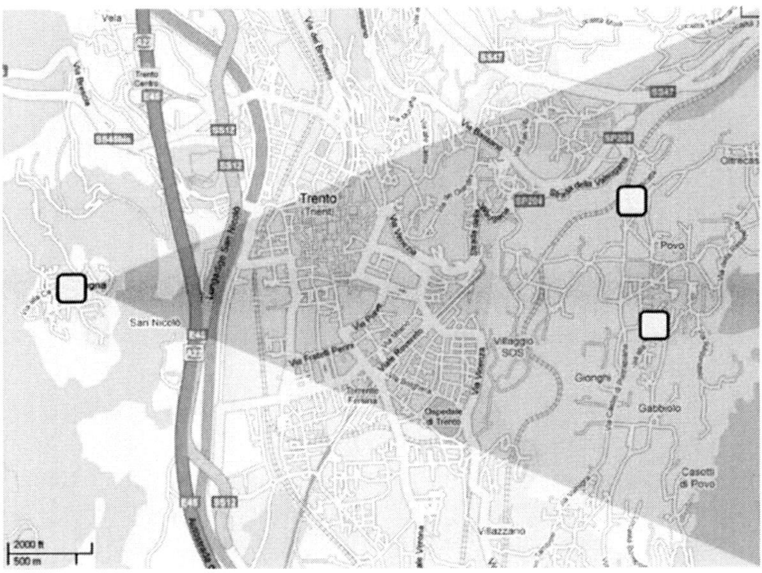

Figure 1. WING/WORLD TESTBED GEOGRAPHICAL DISTRIBUTION.

3.1. Hardware Platform

Mesh routers are built exploiting three different types of processor boards, namely, the PCEngines ALIX 2C2 (500 MHz x86 CPU, 256 MB of RAM) processor board, the PCEngines WRAP 1E (233 MHz x86 CPU, 128 MB of RAM) processor board, and the Gateworks Cambria GW2358-4 (667 MHz ARM CPU, 128 MB of RAM). Operating system and application are stored on a 1 GB Compact Flash card for the PCEngines platforms and on the 32 MB embedded flash memory for the Gateworks platform (in this case a 4 GB Compact Flash is used to provide additional storage). It is worth noting that the full WING/WORLD firmware including development and testing tools (traffic generator, loggers, etc.) requires about 16 MB of storage space. A stripped down version of the firmware without development and testing tools requires less than 4 MB of storage.

Connectivity is provided by 2 Ethernet channels, 2/4 miniPCI slots, (PCEngines/Gateworks) and one serial port. PCEngines ALIX/WRAP boards are equipped with two Mikrotik R52 WiFi IEEE 802.11a/b/g cards based on the Atheros AR2412 chipset. Gateworks boards are equipped with two Ubiquiti SR71-A WiFi IEEE 802.11a/b/g/n cards based on the Atheros AR9160 chipset. On the PCEngines platform, one interface builds and maintains the multihop wireless backhaul, while the other interface can be configured either in Client or in Master mode. The former configuration allows the node to share an already available WiFi connection with the entire WMN while the latter configuration is used to provide a standard IEEE 802.11 Access Point. Single interface setups are also supported; however, in this case the device acts as a pure relay node. Dual and single NIC nodes can coexist in the same network. On the Gateworks platform, both interfaces are used to build and maintain the multihop wireless backhaul implementing a true multiradio/multichannel WMN exploiting dynamic channel assignment. An additional WiFi interface can also be used to provide Internet connectivity to the network. Finally, platforms with USB support (PC Engines ALIX and Gateworks Cambria) can be equipped with a cellular modem allowing the entire WMN to exploit a UMTS/GPRS network as a gateway link to the Internet.

Being based on the x86 architecture, the PCEngines boards (similar systems are provided by Soekris Engineering) deliver high flexibility in terms of choice of components while at the same time providing us with platform suitable for real-world deployments in terms of both maintenance costs and expected performances. Moreover, no cross compilation is required and standard development tools and OSes can be used. On the other hand, the Gateworks boards deliver higher performances and support up to 4miniPCI wireless adapters enabling effective multiradio/multichannel deployments.

The selection of the Wireless NIC to be deployed in our testbed has been driven by the need to configure and control as many low-level (physical) parameters as possible. The selected Atheros-based Wireless NICs allow to control parameters such as transmission bit-rate, carrier sense thresholds from userspace and provide transmission feedback for unicast frames which are not successfully delivered. Moreover, raw 802.11 frames are exposed by the driver, allowing to control most of the node's functionality at the application level; for example, it is possible to get per-packet signal and noise readings and to send broadcast frames at arbitrary rates.

3.2. Software Platform

OpenWRT [16] has been selected as Operating system for our testbed. OpenWRT is a minimalist BusyBox/Linux distribution released under a GPL license [17]. It provides an automated system for downloading the source code for both the kernel and the userspace tools, and compiling it to work on any supported platform. Moreover, it is characterized by a small memory and disk footprint making it suitable for a wide rage of networking devices. Finally, it provides hardware configuration and maintenance abstraction through a custom system and package configuration facility called UCI (Universal Configuration Interface) and exploiting MIB-like structure in order to streamline device management using SNMP [18].

It is worth noting that, being based on the x86 architecture, the PCEngines processor boards do not require, in principle, cross-compilation; however, we decided to use OpenWRT in order to abstract from the underlying hardware architecture making the WING/WORLD toolkit a platform-agnostic solution for wireless mesh networking. As a matter of fact, OpenWRT proved to be the most effective "glue" between heterogeneous hardware and software requirements.

The overall software architecture is sketched in Figure 2. As it can be seen from the picture, the node supports multiple backhauling technologies (Wired, WiFi, and UMTS). The software can seamlessly switch from one backhaul link the the other. However, due to the use of Network Address Translation (NAT) techniques at the mesh gateway, existing connections exploiting stateful protocols, such as TCP, are terminated when the backhaul link is switched. Likewise end-users applications that rely on NAT traversal techniques in order to implement client-to-client communications (e.g., peer-to-peer and VoIP) cannot survive the transition and are bounded to drop their active connections. It is worth underlying that, such a behavior derives from the joint use of NAT and network masquerading (or IP masquerading). This technique allows the network administrator to "hide" an entire address space (typically private network addresses) behind a single IP address (typically a public address). If on one hand such a technique allowed to tackle the exhaustion of IPv4 address space, on the other hand hosts behind NAT-enabled routers do not have end-to-end connectivity, which in time breaks the operations of stateless protocols such as UDP or hinder the services that require the initiation of TCP connections from the outside network.

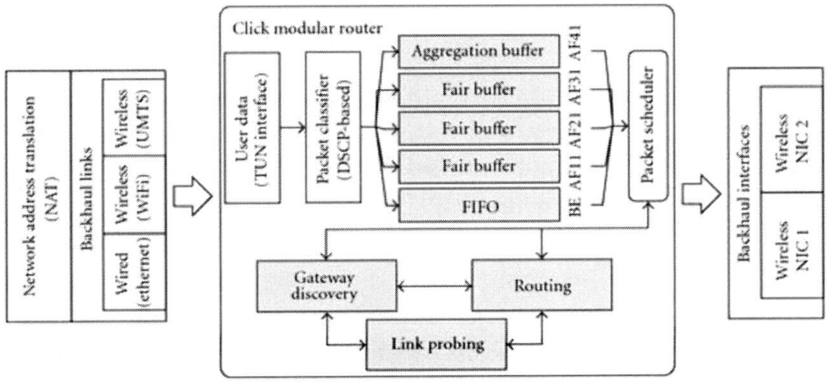

Figure 2. ARCHITECTURE OF THE TRAFFIC DIFFERENTIATION SCHEME IMPLEMENTED IN THE WING/WORLD TOOLKIT.

The above mentioned issues are typically addressed using a variety of NAT traversal protocols. NAT traversal is a generic term used to identify techniques that establish and maintain TCP and/or UDP connections across NAT gateways. The general goal of a client implementing a NAT traversal protocol is to know its own external address (i.e., the address behind which the local address space has be hidden). The client can then start the communication by advertising its external NAT address to its peers, rather than the masqueraded (local) address that is not reachable for its peers on the public network. However, NAT traversal techniques are not designed to handle dynamic network egress points, as a matter of fact, a client has no way of notifying its peers that its external address has changed. Possible solutions involve modifying currently used NAT traversal protocols (e.g., Session Traversal Utilities for NAT [19]) in order to support dynamic network egress points or implementing a Mobile IP architecture in order to allow end-users' client to roam across different public networks. Both approaches are highly invasive and as such are out of the scope of this work.

Routing software is implemented using the Click modular router [20]. A Click router is built by assembling several packet processing modules, called elements, forming a directed graph. Each element is in charge of a specific function such as packet classification, queuing, and interfacing with networking devices. Click comes with an extensive library of elements supporting various types of packet manipulations. Such a library enables easy router configuration by simply choosing the elements used and the connections among them. Finally, a router configuration can be easily extended by writing new elements. The Click modular router is available as both Linux Kernel Module and user-space driver, allowing straightforward porting of a user-space implementation to kernel-space. Mesh routers uses the Click software router toolkit for route/gateway discovery, packet forwarding, and to implement a DiffServ-like traffic differentiation architecture. These features are sketched in Sections 3.3 and 3.4, respectively, while details about the modular gateway architecture (left hand block) and the multiradio mesh backhaul (right hand block) are given in Section 4.

3.3. Routing Framework

The WING/WORLD toolkit is built on top of the Roofnet platform. Roofnet is an experimental WMN developed by the MIT. The Roofnet architecture is described in detail in [21]. Roofnet routes packets using a DSR-like routing protocol called *SRCR* exploiting the Estimated Transmission Time (ETT) as routing metric [22] and optimized for network scalability and throughput rather than for supporting mobility. The ETT metric aims at estimating the amount of time required to transmit a packet over a wireless link (including retransmission). The ETT metric is computed as follows:

$$M_{\mathrm{ETT}} = \frac{1}{P_{\mathrm{ACK}} R}, \tag{1}$$

where R is an estimate of the highest effective throughput achievable in the forward direction, and P_{ACK} is the delivery probability of the ACK signal in the reverse direction (d_{rev}). Since r_x is the estimated throughput of broadcast packets in the forward direction at the transmission rate of x Mb/s, the parameter R can be computed as follows:

$$R = \max(r_1, r_2, r_{5.5}, r_{11}),$$

$$r_x = d_{\mathrm{fwd}} x, \tag{2}$$

where d_{fwd} is the link delivery probability in the forward direction. In order to compute the forward (d_{fwd}) and reverse (d_{rev}) link delivery ratios, each node periodically broadcasts a sequence of five probes: one short probe aimed at modeling the ACK transmission and one long probe for each available transmission rate (broadcast frames are not acknowledged nor retransmitted by IEEE 802.11 devices). Each node keeps track of the number of probes received during an observation window W. At any time, d_{rev} is then given by.

$$d_{\text{rev}}(t) = \frac{\text{count}(t - W, t)}{w/\tau}. \tag{3}$$

Note that $\text{count}\,(t - W, t)$ is the number of probes received during the observation window W, and w/τ is the number of probes that should have been received. Finally each probe sent by a node contains the number of probes packets received by the same node from all its neighbors during the last observation window. Such a design choice allows the receiver to compute the forward delivery ratio d_{fwd} toward the node from which the probe was originated. Using two probes to estimate data and ACK delivery ratios separately allows the routing layer to properly model asymmetric links and to cope with the hidden node phenomena. In fact, probes lost at the receiver side due to interference are taken into account during the computation d_{fwd} at the transmitting side by exploiting the information piggy-backed into each probe.

The default Roofnet implementation has been extended with additional modules responsible for QoS management. These enhancements are described in detail in [23–25]. For readers' convenience, a brief overview of their main features is provided in the next section.

3.4. QoS Extensions

The WING/WORLD toolkit implements a traffic prioritization scheme based on the DiffServ [26] framework in order to allow classification and differentiated treatment. Network traffic entering a mesh router is classified by DSCP code and then fed to a suitable queue. Traffic differentiating is provided by means of a Deficit Weighted Round Robin (DWRR) scheduler which pulls packets from buffers, according to some input weights (see Table 1).

Table 1. PHBs supported by the WING/WORD module for Differentiated Services.

DSCP	PHB	Weights	Queuing	Traffic Type
0	Default	1	ADRR	*Best Effort*
0x0A	AF11	2	ADRR	*Low Priority*
0x12	AF21	4	ADRR	*Medium Priority*
0x1A	AF31	8	ADRR	*Streaming (UDP)*
0x22	AF41	8	ADRR w/A-MSDU	*Real-time (UDP)*

Each buffer maintains a pool of queues and a hash table that associates the MAC destination addresses with one of those queues. Incoming MAC frames are first classified according to their destination address and then fed to the corresponding queue. If such a queue does not yet exist, it is created dynamically by the scheduler. Unused queues are moved from the hash table to the pool. This is done in order to alleviate the need for repeated memory allocation as neighbors come and go. Within each buffer, two different link scheduling policies are supported by the system:

i. *Airtime Deficit Round Robin (ADRR)*. It aims at providing intracell *airtime* fairness. ADRR enhances the Deficit Round Robin (DRR) discipline by taking into account the channel quality which in time prevents a node affected by high-packet loss from monopolizing the wireless channels thus lowering the performance of the whole system.

ii. *ADRR with Frame Aggregation*. It aims at reducing the MAC service time by concatenating several MAC Service Data Units (MSDUs) to form the data payload of a large Aggregated-MSDU (A-MSDU). Such packet aggregation scheme leverages the channel probing functionalities of mesh routers in order to compute the optimal saturation burst length.

4. PLATFORM DETAILS AND OPTIONS

This section provides additional details on the features currently supported by the WING/WORLD toolkit. The system components illustrated in the following subsections are the following:

(i) self-configuring backhaul,

(ii) multipleradio support,

(iii) opportunistic scheduling,

(iv) traffic aggregation,

(v) authentication.

However, given the modular nature of the platform, it must be underlined that additional modifications or extensions can be easily introduced. As a matter of fact, each of the aforementioned components is independent from the underlying routing layer and can be readily used in conjunction with other routing protocols implementations (i.e., OLSRd [27], BATMAN [28], etc.). For example, both the authentication and the self-configuring backhaul are implemented using standard tools available on any Unix-like platform (GNU/Linux, all the children of BSD, etc.). Likewise both the opportunistic scheduling and traffic aggregation modules do not break the standard ISO/OSI (with the exception of the cross-layer interfaces used to access link level parameters; however, such interfaces can be easily adapted to other link-aware routing protocol) layering allowing straightforward porting to other platforms and routing protocols.

4.1. Self-Configuring/Self-Healing Backhaul

The presence of a backhaul technology represents one key differentiation points between WMN and the traditional MANET paradigm. The term *backhaul* is generally used [2] to identify a technology in charge of forwarding the traffic from the originator node to an external network (i.e., the Internet). The WING/WORLD nodes can automatically detect if they are relays or mesh gateways. A mesh node autoconfigures itself as gateway if an IP address can be obtained using DHCP over one of its backhaul links and if a list of well-known Internet addresses can be reached. At the present moment, three different backhaul technologies are supported.

i. *Wired*. The first Ethernet interface available on the node, typically the *eth0* device.

ii. *Wireless (WiFi)*. In dual radio setups, the second WiFi interface can be configured in *Client* mode allowing the node to exploit an existing IEEE 802.11 AP as backhaul link to the Internet.

iii. *Wireless (Cellular)*. If a cellular modem is available, the mesh node can exploit an UMTS/GPRS network as backhaul link to the Internet. The cellular modem must be equipped with an SIM card holder (i.e., Huawei E169, Sierra 881, etc.) and an active SIM card must be inserted.

In the current implementation, only one backhaul link can be active at a given time; however, different mesh gateways can exploit different backhauling technologies providing the testbed with an increased resiliency to network failures. It is worth noting that using more than one backhauling technology at the same time would not increase the network performances in that the typical bottleneck in a WMN lies at the last hop toward the mesh gateway whose capacity is at least an order of magnitude smaller than any of the available backhauling technology (with the sole exception of the UMTS backhaul which is to be considered anyway as a backup solution).

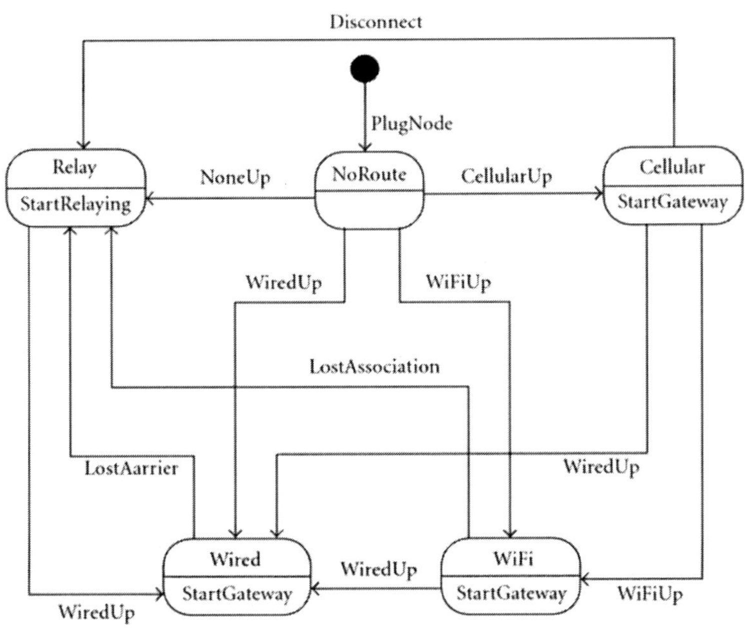

Figure 3. STATE DIAGRAM FOR THE GATEWAY MODULE. Events that cannot occur in a given state are not accounted.

The wired interface takes precedence over both wireless technologies, while the WiFi backhaul takes precedence over the cellular link. A Finite State Machine (FSM) has been implemented in order to properly handle the backhaul's configuration without having either to reboot the node or to disrupt the normal network operations (Figure 3). Each backhaul technology is associated with a state. Additionally, the FSM defines the following states.

i. *Relay*. None of the backhaul links are active. The node configures itself as pure relay node. In dual-radio setups, the node also act as IEEE 802.11 AP providing the end-users with standard hotspot.

ii. *NoRoute*. None of the backhaul links are active and no multihop mesh backhaul could be configured. The node configures itself as an IEEE 802.11 AP; however no Internet connectivity is provided to the hotspot.

The specification of the FSM is provided in Table 2 as State Transition Table (STT). The vertical dimension indicates the current state; the horizontal dimension indicates possible events; the row/column intersections contains the next state if an event happens and the actions to be performed when the state transition occurs. Please note that *PlugNode* is an event that comes from the external world (i.e., powering up the mesh node). Here follows a description of all the possible events.

Table 2. State transition table. Events that cannot occur in a given state are not accounted.

Current state	Event	Action	Next state
null	PlugNode	—	*NoRoute*
NoRoute	WiredUp	*Start Gateway*	*Wired*
NoRoute	WiFiUp∧¬WiredUp	*Start Gateway*	*WiFi*
NoRoute	CellularUp∧¬WiFiUp∧¬WiredUp	*Start Gateway*	*Cellular*
NoRoute	NoneUp	*Star Relaying*	*Relay*
Wired	LostCarrier	*Star Relaying*	*Relay*
WiFi	WiredUp	—	*Wired*
WiFi	LostAssociation	*Star Relaying*	*Relay*
UMTS	WiredUp	—	*Wired*
UMTS	WiFiUp	—	*WiFi*
UMTS	Disconnect	*Star Relaying*	*Relay*
Relay	WiredUp	*Start Gateway*	*Wired*
Relay	WiFiUp∧¬WiredUp	*Start Gateway*	*Wireless*
Relay	CellularUp∧¬WiFiUp∧¬WiredUp	*Start Gateway*	*Cellular*
Relay	NoneUp	—	*Relay*

i. *WiredUp*. The node successfully obtained an IP Address over its wired interface.

ii. *WiFiUp*. The node failed to obtain an IP Address over its wired interface, however it succeeded in associating an authenticating with a pre-configured IEEE 802.11 AP.

iii. *CellularUp*. The node failed to obtain an IP Address over both its wired and its wireless interfaces; however, it succeeded in establishing a direct connection with the UMTS/GPRS network using the Point-to-Point Protocol (PPP).

iv. *LostAssociation*. The wireless interface has lost its association with the AP being used as wireless backhaul. This event can occur only if the node is in the *Wireless* state.

v. *LostCarrier*. The wired interface has lost its carrier on the wired interface. This event can occur only if the node is in the *Wired* state.

vi. *Disconnect*. The PPP connection has been terminated by either of the parties involved in the communication.

vii. *NoneUp*. The node failed to activate any of the supported backhaul links.

Two distinct actions can be linked to a state transition: *Start Gateway* and *Start Relaying*. The former action sets the backhaul link associated with the new state as the default route to the Internet; moreover, the node starts advertising itself as candidate gateway for the mesh network. The latter action disables the current backhaul link and use the multihop wireless backhaul as default route to the Internet.

The performances of the self-configuring and self-healing backhauling technology have been assessed through a series of experimental tests aimed at modeling different kinds of failures that can happen at a WMN's gateway. The reference network configuration is sketched in Figure 4. Internet connectivity is provided to the WMN by the *New Delhi* mesh gateway. The *Rome* desktop provides services such as DHCP, Radius, and PBX (Private Branch exchange). Finally, the *Quito* laptop acts as end-user's device exploiting the standard WiFi Hotspot provided by the *Alix 6* mesh router. The network implements a three-tiers architecture [2] where traffic generated my end-users (first tier) is relayed to the destination by the mesh backbone (second tier). Internet-working with external networks (i.e., the

Internet) is provided by dedicated nodes called mesh gateway that can exploit aforementioned backhauling technologies (third tier).

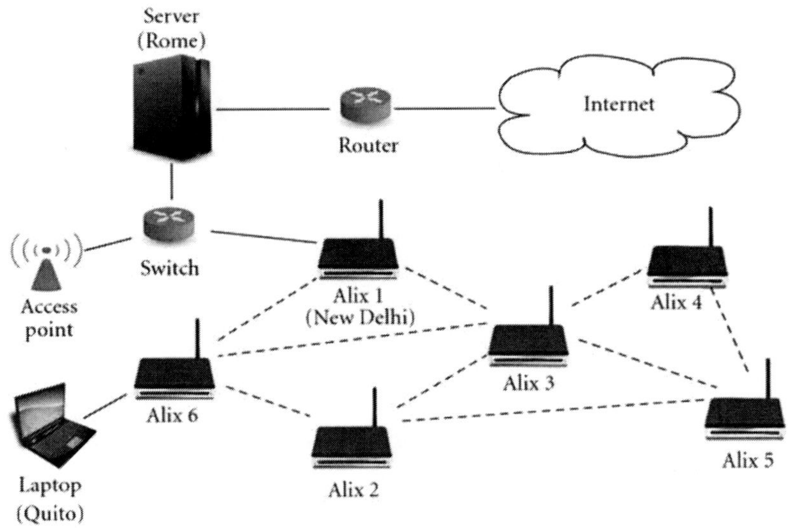

Figure 4. NETWORK CONFIGURATION USED TO ASSESS THE PERFORMANCES OF THE PERFORMANCES OF THE SELF-CONFIGURING AND SELF-HEALING BACKHAULING TECHNOLOGIES.

The target of the measurement campaign carried over the WING/WORLD testbed aimed at evaluating the time spent by mesh gateway to switch to another backhauling technology when the current one is experiencing service outages as well as to revert to the original configuration when connectivity has been restored. In order to do so, the following steps have been undertaken.

- (1)The mesh gateway is connected to the Internet using a wired connection (Ethernet).
- (2)The *Quito* wireless client associates to the WiFi Hotspot provided by an the *Alix 6* mesh router.
- (3)The wired connection is made unavailable by yanking the Ethernet cable from the mesh gateway. This step triggers the transition *Wired →Wireless*.

- (4)The wireless connection is made unavailable by turning off the WiFi Access Point. This step triggers the transition *Wireless →UMTS.*
- (5)The wireless connection is restored. This step triggers the transition *UMTS →Wireless.*
- (6)The wired connection is restored. This step triggers the transition *Wireless →Wired.*

Switching time has been evaluated in two different scenarios. In the former one, end-users' Internet connectivity has been assessed by continuously pinging a remote host (http://www.google.com/) from *Quito*. Ping period has been set to 200 ms. In the second scenario, a synthetic traffic flow has been generated from *Quito* to *Rome*. The traffic flow has been modeled according to the parameters of the G.729.3 codec [29], a widely used VoIP codec. The G.729.3 VoIP codec generates 33 pkts/s; each packet contains 3 voice samples (10 bytes each) producing a final bit-rate of 8 kbits/s. All measurements are averaged over 10 runs.

The results of the first scenario are reported in Figure 5. The average Round-Trip-Times (RTTs), estimated using the *ping* command, are 50.2 ms, 51.7 ms, and 167.8 ms, respectively, for the Ethernet, WiFi, and UMTS backhauls. Confidence intervals are smaller than 2 ms over either backhaul links.

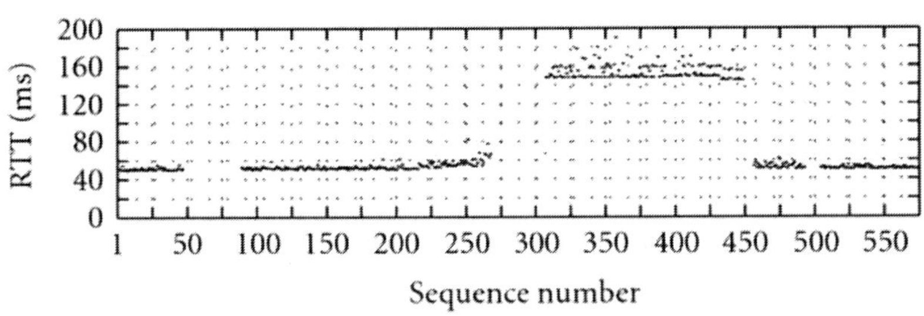

Figure 5. SEQUENCE NUMBER (SN) VERSUS ROUND-TRIP-TIME (RTT) AS REPORTED BY THE ping COMMAND. Each point represents a single RTT value. This graph shows the *Ethernet/Wifi* transition (SN between 120 and 220), the *WiFi/UMTS* transition (SN between 270 and 305), the *UMTS/WiFi* transition (SN between 570 and 620), and the *WiFi/Ethernet* transition (SN between 770 and 780).

The results of the second scenario are reported in Figure 6 and summarized in Table 3. As it can be seen from the picture, a transition that is the result of a service disruption (*Ethernet →Wireless*, *Wireless →UMTS*, and *Ethernet →UMTS*) requires a longer switching time than a transition that is reverting the backhaul, to its original configuration. The reason for this behavior lies in the fact that in the former case, about 6 seconds are spent by the mesh gateway to negotiate its IP address with the DHCP server. Moreover, in the transition to the UMTS backhaul a considerable amount of time is spent in the PPP negotiation phase with the UMTS carrier. On the other hand, in restoring a backhaul link, the mesh gateway can first complete the IP negotiation phase and then it can switch its default route to the new link.

Table 3. Average backhaul switching times estimated using a synthetic UDP stream. Confidence intervals (95%) are shown in parenthesis. Results are in seconds.

From → To	Ethernet	WiFi	UMTS
Ethernet	—	10.6 (1.3)	25.9 (1.5)
WiFi	2.3 (0.4)	—	25.8 (1.9)
UMTS	2.8 (0.4)	9 (0.2)	—

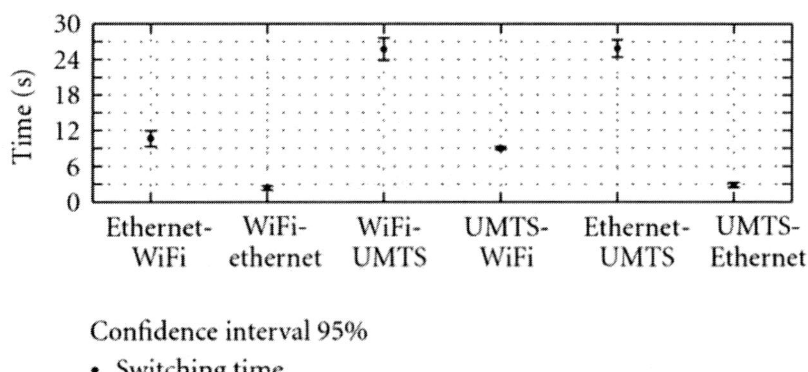

Figure 6. BACKHAUL SWITCHING TIME ESTIMATED USING A SYNTHETIC UDP STREAM MODELED ACCORDING THE PARAMETER OF THE G. 729.3 VoIP codec. This graph shows the backhaul switching time for all the possible transition. Each point is on average of 10 runs.

4.2. Multiradio

The WING/WORLD toolkit can exploit multiple radios per mesh router, allowing simultaneous transmissions and reducing intrapath interference by tuning the mesh radios on non-overlapping channels. The Interference and Traffic Aware Channel Assignment (ITACA) algorithm has been developed in order to both assign the channels efficiently by taking into account the effects of traffic and interference patterns and to maintain topological connectivity. ITACA uses the Channel Assignment Server (CAS), which can be colocated with the mesh gateways, as a central node to collect information from the network and to assign channels to each radio interfaces. The objective of the CAS is to minimize the interference between mesh routers, and also to minimize the interference between the mesh network and other colocated wireless networks. It adopts a hybrid approach in assigning channels, combining pseudostatic default channel assignment, and dynamic channel assignment. It is worth noting that our approach ensures that channel assignment does not alter the network topology by mandating that one radio on each mesh router must operate on a default channel.

ITACA assigns channels considering both interference and traffic distribution. When traffic is homogeneously distributed among all nodes, ITACA assigns channels starting from the gateway, selecting links with the best metric. This approach is not optimal in case of traffic that is not homogeneously distributed among all nodes. In order to address such case, we consider the coefficient of variation of the aggregated traffic crossing each link. If this coefficient is greater than a threshold (80% in our implementation), we give higher priority to links transmitting more data while assigning channels. Otherwise, if the coefficient is smaller than the threshold, our scheme sorts links considering only interference information, thus giving higher priority to links emanating from the gateway and going toward the edges of the network. A multiradio conflict graph model [30] is used to estimate and model the interference within the network and also between the network and other colocated wireless transmitters. The Coefficient of Variation (c_v) is defined as the ratio between the standard deviation (σ) and the mean value (μ):

$$c_5 = \frac{\sigma}{\mu}, \tag{4}$$

where mean value (μ) and standard deviation (σ) are respectively defined as:

$$\mu = \frac{1}{N}\sum_{i=1}^{N} x_i, \qquad \sigma = \sqrt{\frac{1}{N}\sum_{i=1}^{N}(x_i - \mu)^2}. \tag{5}$$

The ITACA algorithm is being currently evaluated over the WING testbed to assess performance and the effectiveness of the channel assignment strategy with respect to the operation of the mesh network. The ITACA algorithm has already been implemented and tested using the NS-2 simulator [31]. Figure 7 shows the observed performance improvement of ITACA with respect to a dynamic channel assignment algorithm, BSF-CA [32]. Results are averaged over 10runs. From the figure, it can be observed that ITACA outperforms BSF-CA scheme with respect to throughput and channel assignment delays. The simulations were carried out considering the presence of external network interference. The simulation details and the network models are presented in [30]. The multiradio extensions and the ITACA algorithm, implemented in NS-2, are available for public use at the WING community website [8].

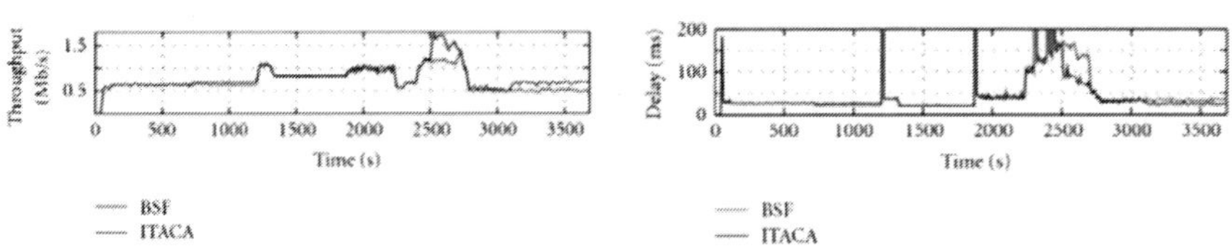

Figure 7. PERFORMANCE IMPROVEMENT OF ITACA WITH RESPECT TO THE DYNAMIC CHANNEL ASSIGNMENT ALGORITHM BSF-CA. Aggregated throughputEnd-to-end delay

4.3. Opportunistic Scheduling

WMNs are known to be susceptible to the "IEEE 802.11 performance anomaly" [33] which refers to sudden performance drops that occurs when nodes transmitting at low bit-rates due to poor channel conditions capture the wireless medium for longer periods of time at the expense of nodes transmitting at higher bit-rates. Intracell airtime fairness is provided in the WING toolkit by the Airtime Deficit Round Robin (ADRR) link scheduling discipline. ADRR enhances the Deficit Round Robin (DRR) discipline by taking into account the channel quality which in time prevents a node affected by high-packet loss from monopolizing the wireless channels thus lowering the performance of the whole system. It is worth underlining that ADRR does not require modification to the standard IEEE 802.11 MAC and can be readily implemented using off-the-shelf components. Figure 8 depicts the main building block of the ADRR scheduler.

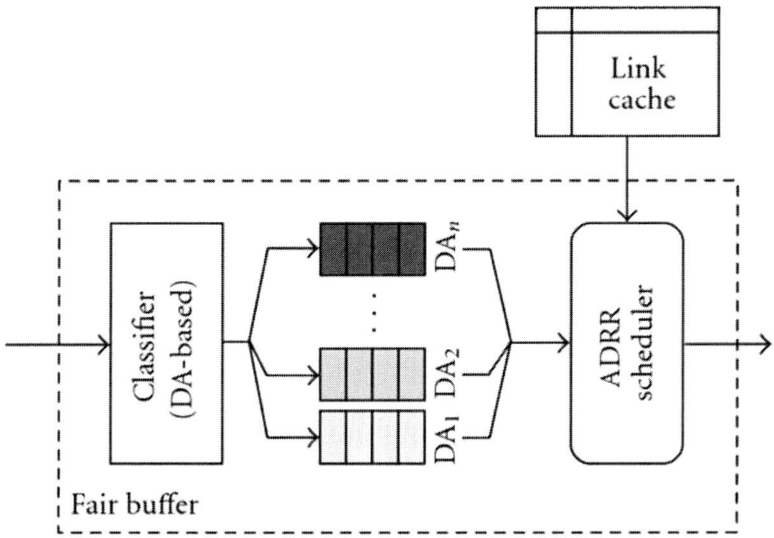

Figure 8. BLOCK DIAGRAM FOR AIRTIME DRR SCHEDULER.

The scheduler maintains a linked list of currently backlogged queues. Each queue is associated with a counter, called Deficit Counter, that indicates the amount of resources the link can use in a round. At each round, the deficit counter of the currently visited queue is increased by a fixed quantity called Quantum. The ADRR scheduler only serves packets whose expected transmission time is smaller than the deficit counter. The expected transmission airtime $TX_{\text{AIRTIME}}$for a packet Sbytes long is given by.

$$TX_{\text{AIRTIME}} = M_{\text{ETT}} \frac{S}{L_{\text{Probe}}}, \tag{6}$$

where $M_{\text{ETT}}$is the the link metric and $L_{\text{Probe}}$is the size of the probe used to compute it. Basically the transmission airtime is the link's metric linearly scaled in order to take into account the size of the packet being transmitted. As a matter of fact, given a certain bit error rate and assuming the errors are i.i.d. (after decoding), a longer frame has a higher probability of getting corrupted and thus will require a longer transmission time. After a packet is sent, the deficit counter is decreased by the expected transmission time of the transmitted packet. A frame whose transmission time exceed the deficit counter is held back until the next visit of the scheduler. Empty queues are removed from list of currently backlogged queues and their deficit counter is set to zero.

Measurements carried out over the WING/WORLD testbed proved the capability of the ADRR scheduler to provide performance isolation in IEEE 802.11-based WMNs. Network topology is sketched in Figure 9(a). In particular, nodes number 2, 3, and 4 are fed with a CBR connection generated at node number 1. We have modeled each CBR connection as a single UDP stream with constant inter-departure time (2 ms) and packet size (1460 bytes) producing a final bit-rate of ≈ 6 Mb/s.

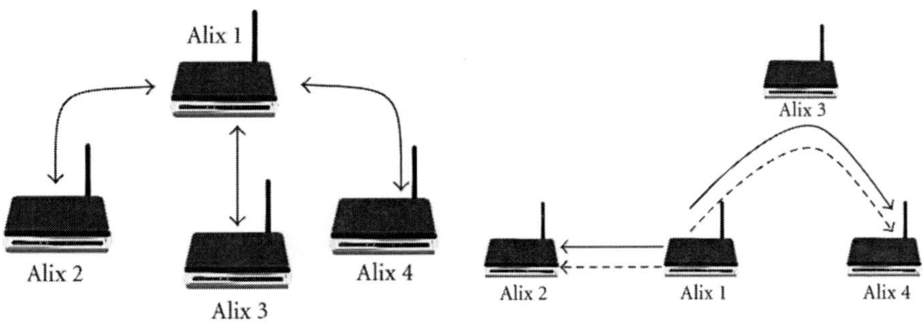

Figure 9. NETWORK TOPOLOGIES EXPLOITED DURING THE MEASUREMENT CAMPAIGN. Star-shapedString

Measurements have been carried out exploiting three deployments scenarios differentiated by the channel condition experienced by nodes number 2. Notice that in each deployment all nodes are in radio range. However, while node number 3and 4are kept close to the gateway, node number 2is positioned in such a way to experience channel condition raging from *Good* to *Poor* with an intermediate *Medium* quality.

Tables 4 through 6 summarize the outcomes of our measurements campaign. Results show that the proposed scheduler is capable of addressing the *"IEEE 802.11 performance anomaly"* maintaining a high throughput over reliable links (Node number 3 and 4) as opposed to both the FCFS and the DRR policies that fail to achieve performance isolation when node number 2 starts to experience poor channel conditions.

Table 4. Average throughput for different scheduling disciplines (Good channel conditions). Results are in Kb/s.

Scheduler	Node 2	Node 3	Node 4	Aggregated
FCFS	1869.6	1818.2	1870.7	5558.5
DRR	1857.1	1817.8	1889.2	5564.2
ADRR	2014.2	1995.4	2055.9	6065.5

As is shown in Table 4, when channel condition for node number 2 is still good, the available resources are evenly shared among all the nodes. However, it is worth noting that the average throughput achieved by each node using the ADRR is slightly higher than the throughput achieved using both the FCFS and the DRR scheduling disciplines.

We postulate that the ADRR scheduler is capable of exploiting channel fluctuation by opportunistically allocating more airtime to links that experience better channel condition. We recall that the feedback mechanism embedded in the routing metric gives the transmitting station (Node 1 in our case) the capability to schedule for transmission links experiencing better channel conditions. Such considerations are supported by the theoretical finding in [34] where channel fluctuations are exploited by transmitting information opportunistically when and where the channel is strong.

As node number 2 moves away from the gateway (Tables 5 and 6), the ADRR is capable of allocating more resource to the nodes experiencing better channel conditions, while the other scheduling policies degrade the aggregated throughput. In the extreme case where node number 2experience poor channel condition, ADRR outperforms both FCFS and DRR by delivering a higher aggregated throughput (1.1 Mb/s w.r.t. the baseline scenario) and by allocating to nodes number 3 and 4 a percentage of the bandwidth which is only slightly lower than the optimal case.

Table 5. Average throughput for different scheduling disciplines (Medium channel conditions). Results are in Kb/s.

Scheduler	Node 2	Node 3	Node 4	Aggregated
FCFS	1626.2	1579.1	1616.5	4821.9
DRR	1751.0	1709.6	1785.3	5245.9
ADRR	1751.6	2024.8	2074.2	5850.6

Table 6. Average throughput for different scheduling disciplines (Poor channel conditions). Results are in Kb/s.

Scheduler	Node 2	Node 3	Node 4	Aggregated
FCFS	1364.8	1318.7	1298.4	3981.9
DRR	1399.1	1340.9	1323.3	4063.3
ADRR	1107.2	1993.6	2038.8	5139.6

4.4. Traffic Aggregation

In order to increase the performances of the WING testbed, we implemented a packet aggregation technique capable of reducing the overhead due to both protocol headers and the channel contention by concatenating several MAC Service Data Units (MSDUs) to form the data payload of a large MAC Protocol Data Unit (MPDU). Such packet aggregation scheme leverages the channel probing functionalities of mesh routers in order to compute the optimal saturation burst length. The building blocks of the Aggregation Buffer and their relationships are sketched in Figure 10.

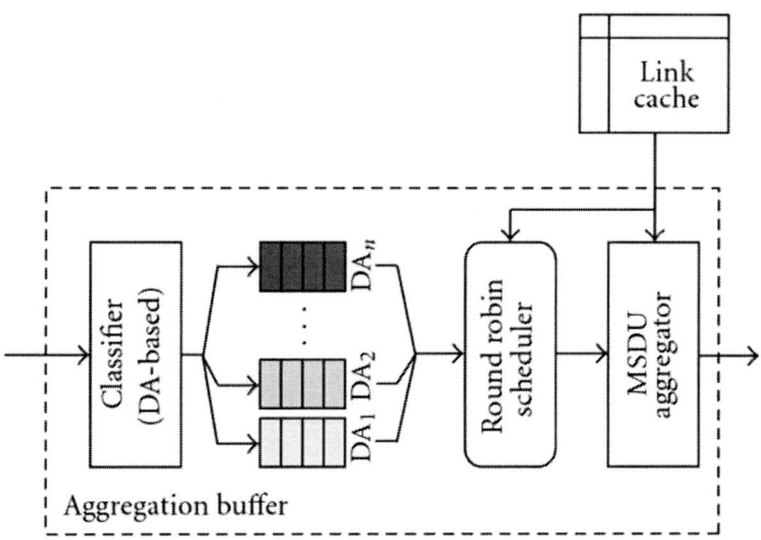

Figure 10. BLOCK DIAGRAM FOR AGGREGATION BUFFER.

Incoming MAC frames are first classified according to their destination address and then fed to a different queue. For each queue, an A-MSDU is generated when either an aggregation timer is expired or a burst of optimal length can be generated. The ETT metric is exploited as a cross-layer technique in order to match link layer parameters with our adaptive traffic aggregation policy.

Aggregation and deaggregation are performed at each hop. Albeit such an approach could lead to increasing delays as the number of hops increases, we postulate that, at intermediate nodes, medium access delay is sufficient to collect enough packets so that burst generation is triggered by the optimal frame length without incurring in any aggregation delay. In our measurement settings, each mesh node sustains the same traffic, consisting in an increasing number of VoIP sessions plus an additional background traffic modeled according to a TCP socket working in saturation regime. Streams configuration is sketched in Figure 9(b) where VoIP bundles and TCP flows are represented respectively by dashed and solid lines. Each VoIP call has been emulated as single UDP stream modeled according to the parameters of the G.729.3 codec [29]. A typical VoIP source tends to transmit a large number of packets with a small payload, and such a combination is known to lead to large transmission overheads [35] in that a considerable amount of time is wasted in the contention phase and in sending headers and acknowledgments.

The outcomes of the measurements campaign are reported in Figure 11. As it can be seen from the figure, our adaptive aggregation policy provides almost a factor 3performance increase, since the number of sustained sessions reaches 28, whereas the plain IEEE 802.11 protocol allows for just 8-9 VoIP sessions.

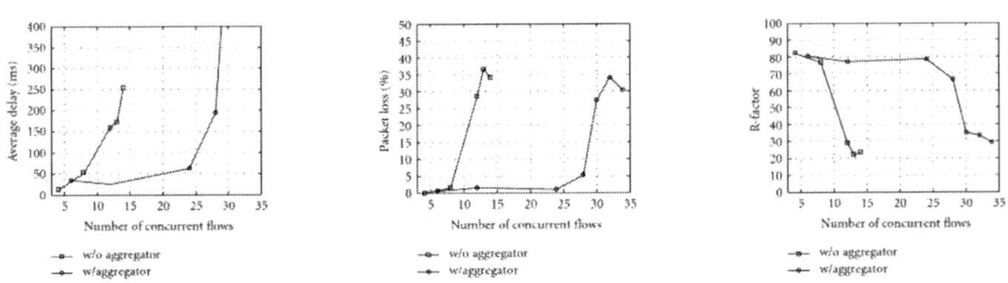

Figure 11. PERFORMANCE MEASUREMENTS RESULTS WITH BACKGROUND INTERFERENCE. Average delay versus number of concurrent flowsAverage packet loss versus number of concurrent flowsAverage R-Factor versus number of concurrent flows

4.5. Authentication, Authorization, and Accounting

End-users access to the WING/WORLD testbed is performed through a captive portal. Captive portals leverage a common web-browser as a secure authentication device, which in time delivers service providers with a standard mean to extend their hotspot coverage using WMN technology while maintaining well-established Authentication, Authorization, and Accounting (AAA) practices.

A captive portal forces an HTTP client on a generic network to see an authentication page before using the Internet normally. In order to achieve such a goal, a captive portal must intercept all packets, regardless of address or port, until the user opens a browser and tries to access the Internet. At that time the browser is redirected to an authentication page. The captive portal authenticates the end-user using against a centralized users directory using a suitable protocol. In our case, the RADIUS [36] protocol has been chosen to provide centralized AAA management for end-users. The main building blocks that compose a captive-portal-based end-users access management system are shown in Figure 12.

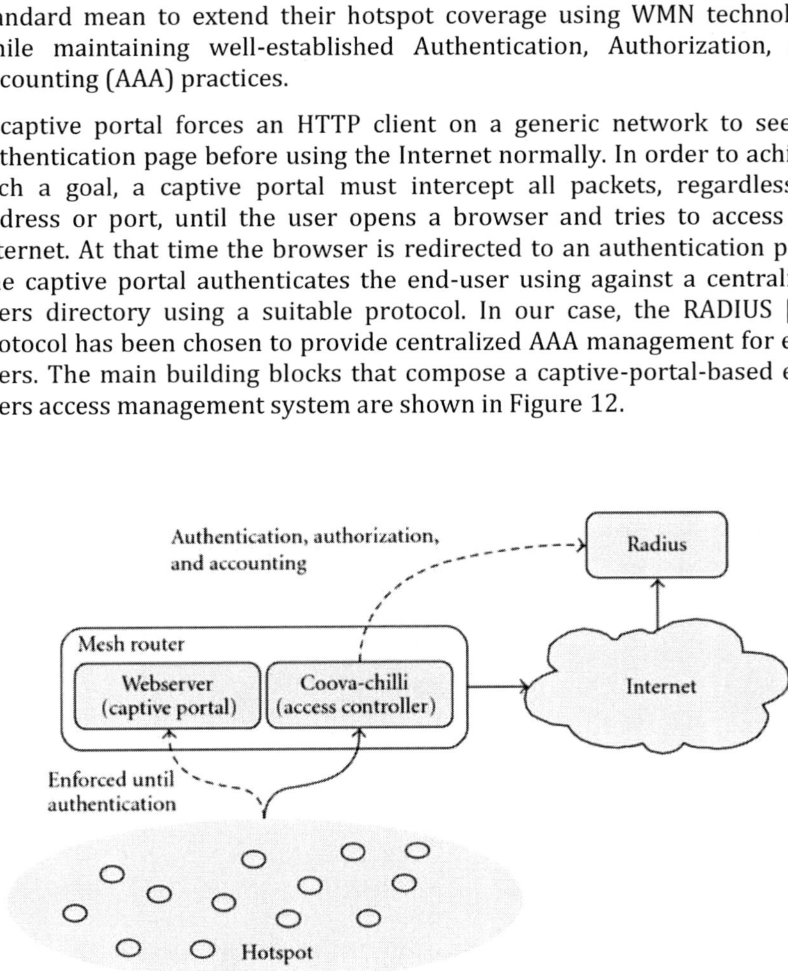

Figure 12. BUILDING BLOCKS OF THE CAPTIVE-PORTAL BASED END-USERS ACCESS MANAGEMENT SYSTEM. Solid lines represent physical communication path, while dashed lines represent logic interaction among the different components.

As the result of a successful authentication the captive portal will update the users directory with the details about the newly authenticated user (MAC/IP address, login time, etc.) and will grant its IP address with the right to access the Internet. Additionally, the captive portal can also set QoS routing rules so that they get provisioned a certain amount of bandwidth or cap to the amount of data transfered.

CoovaChilli [37] has been chosen as wireless LAN access point controller. It supports web-based login (through a captive portal), which is today's standard for public HotSpots, WISP "smart-client" authentication, and it supports Wi-Fi Protected Access (WPA and WPA2). FreeRADIUS [38] is a high-performance and highly configurable RADIUS server. It supports many database back-ends ranging from flat-text files, up to RDBMS and LDAP servers. It also supports many authentication protocols such as PAP, CHAP, and so forth.

It must be underlined at this point that the opportunity to provide access to a general audience people to the WMN infrastructure is a critical point for a testbed, since it enables to offer a realistic load to the network and to capture feedback on the users' perceived quality of the considered architecture (this last version currently under implementation).

As the result of a successful authentication, the captive portal will update the users directory with the details about the newly authenticated user (MAC/IP address, login time, etc.) and will grant its IP address with the right to access the Internet. Additionally, the captive portal can also set QoS routing rules so that they get provisioned a certain amount of bandwidth or cap to the amount of data transferred.

5. COMPARISON WITH OTHER ARCHITECTURES

In this section, we briefly survey some of the most interesting multiradio testbeds currently available in the WMN scene. It is not the authors' intention to provide and exhaustive coverage of all the academic and industrial efforts in this area, instead, we concentrate on solution based on off-the-shelf components and exploiting open-source software. We decided to focus only on multiradio solutions in that, albeit a considerable number of prototypes have been developed and deployed by both the academic and the industrial worlds, little efforts have been dedicated to implement and deploy a multiradio solutions for WMNs. The vendors which have been selling wireless mesh solutions of course do implement some of form multichannel architecture, but they are obviously very reluctant to release those information. Hence the research in WMNs field lacks from a comprehensive perspective a realistic architecture for distributed channel assignment in multiradio WMNs. The most relevant implementations of IEEE 802.11-based multiradio/multichannel WMNs are summarized in Table 7. For the considered WING/WORLD toolkit, two columns are provided: the first related to the version used at the time of this writing in our deployment, and the second related to the next release currently under development.

Table 7. Comparison between different IEEE 802.11-based multiradio/multichannel WMNs implementations.

	Hyacinth	MCL/DCA-MCL	DMesh	ROMA	WING/WORLD	WING/WORLD 2.0
Project Type	Academic	Academic	Academic	Academic	Academic	—
License	GPL/BSD	MSR-SSLA	n.a.	n.a.	BSD	—
OS	GNU/Linux	MS Windows XP	GNU/Linux	GNU/Linux	GNU/Linux	—
Routing Protocol	Spanning tree	MR-LQSR	DOLSR	ROMA	SRCR	MR-SRCR
Routing Metric	Hop-count, Residual Link Capacity	WCETT	ETT	SIM	ETT	WCETT
Management	No	No	No	No	Web-based management dashboard	Decentralized network monitoring and management
QoS support	No	No	No	No	Yes (DiffServ)	Yes (DiffServ)
Users management	No	No	No	No	WPA-PSK, Access Controller (CoovaChilli)*	—

*An external RADIUS server is required for full Authentication, Authorization, and Account (AAA) services.

5.1. MCL

The Mesh Connectivity Layer (MCL) is an experimental Microsoft Windows driver developed by Microsoft Research and released under a Shared Source License. MCL implements an interposition layer between layer 2 (the link layer) and layer 3 (the network layer) of the standard ISO/OSI model. It is sometimes referred to as layer 2.5. To the higher layers, MCL appears to be just another Ethernet link, albeit a virtual one. To the lower layers, MCL appears to be just another protocol running over the physical link. MCL routes using a modified version of DSR [39] called MultiRadio Link Quality Source Routing (MR-LQSR) [22]. LQSR assigns a weight to each link. This weight is the expected amount of time it would take to successfully transmit a packet of some fixed size on that link. In addition, the channel, the bandwidth, and the loss rate are determined for every possible link. This information is sent to all the nodes. Based on this information, LQSR uses a routing metric called Weighted Cumulative

Expected Transmission Time (WCETT) to define the best path for the transmission of data from a given source to a given destination. Channel assignment is static and must be performed manually by the network designer.

An extension to MCL, featuring distributed channel assignment (DCA-MCL in the comparison table), is proposed in [40]. The distributed channel assignment scheme proposed by the authors selects channels that are least used by each node's neighbours. No common channel is used in order to keep the network connected. The algorithm does not need a common signaling channel to keep the network connection. The protocol has been implemented and tested over a 14-nodes testbed. Routes are selected using the MR-LQSR routing protocol using the WCETT metric, a routing metric designed to select channel diverse path in multiradio environments.

5.2. Hyacint

Hyacinth is a multichannel WMN developed by the Experimental Computer System Lab at the Stony Brook University (New York) and built using off-the-shelf components. Each Hyacinth node is equipped with multiple IEEE 802.11 radios operating in ad hoc mode. Internetworking with mobile stations is made possible by a traffic aggregation device embedded in each mesh node. Each WMN nodes interacts with individual mobile stations through the traffic aggregation device and is responsible of assigning mesh-wide unique IP addresses to the mobile stations. A joint centralized channel assignment and routing algorithm has been implemented in order to build and maintain a network spanning tree rooted at each gateway in the network. The protocol does not rely on a common signaling channel to keep the network connected. Hyacinth has been implemented and tested over a small-scale indoor testbed consisting of 9 nodes.

5.3. Mcube

Mcube [41] is a modular, multiradio, WMN designed and developed by the Mobility Management and Networking Laboratory (MOMENT Lab) at UC Santa Barbara. Mcube shows a modular wireless mesh router architecture called *split wireless router*. A *split wireless router* is composed of multiple distinct processing nodes, each equipped with a single radio. Such a design allows for easily extensible wireless mesh routers and alleviates the detrimental self-interference problems that can occur between commodity radios. Channel assignment is performed in a centralized manner; more specifically information about network topology are first collected then the topology and interference aware channel assignment algorithm (TIC) is executed in a central server, and finally the channel assignments are disseminated to the mesh routers. The proposed architecture has been validated using a 20-nodes WMN consisting of 46 IEEE 802.11a/g radios. Results show that an 802.11a dual-radio split router is able to forward aggregate TCP traffic over 15 Mbps. In contrast, a single-unit multiradio router is able to operate at only 2 Mbps because of inter-radio interference. Compared to dual channel assignment schemes (e.g., Microsoft's MCL), TIC's channel selection technique delivers TCP performance improvement in 30%–100% range.

5.4. DMesh

DMesh [12] is an extension to MAP (Mesh@Purdue) [42], a WMN testbed developed and deployed by Purdue University. DMesh exploits both directional antennas for spatial separation and multiple orthogonal channels for frequency separation to provide significantly increased throughput. The Directional OLSR (DOLSR) routing protocol has been developed along with a channel assignment algorithm in order to take advantage of directional antennas setup. DOLSR extends the OLSR protocol assisting the physical formation of high throughput routing trees rooted at the gateways using practical directional antennas, sets up and maintain corresponding routing state, and performs distributed channel assignment. The proposed architecture has been evaluated using both simulation and experiments ran over a mesh network testbed. Results show that, compared with the omnidirectional/multichannel configuration, the proposed architecture improves packet delivery ratio and throughput and drastically lowers average per-packet delay.

5.5. ROMA

ROMA is a joint distributed channel assignment and routing scheme developed by Networking and Wide-Area Systems Group (NEWS) at the New York University. Channel assignment is performed by the network's gateway that broadcast a channel sequence to the other nodes. Channel allocation sequences are computed in such a way that intrapath interference is eliminated. Moreover, ROMA included a measurement driven routing metric inspired by mETX [43] and taking into account link delivery ratio, fluctuations, and external interference. The protocol has been tested on a 24-node dual-radio testbed. Results show that ROMA can achieve high end-to-end throughput and adapts well to changing network conditions.

6. USE CASES

In this section, we shall identify and describe a set of potential use cases for the WING/WORLD toolkit. It is not the authors' intention to survey all the possible application scenarios for WMNs, for such a topic the reader is redirected to [1], instead we aim at providing a set of guidelines for designing and deploying a WING/WORLD-based mesh network.

Unlike the other academic testbeds and prototypes, the WING/WORLD toolkit provides a comprehensive wireless mesh networking solution capable of supporting production level deployments as well as experimentally driven research activities. Moreover, unlike commercial solutions, which are typically monolithic with limited customization possibilities, the "open" philosophy that characterizes the WING/WORLD project empowers researcher and pratictiones with a powerfull tool for implementing and testing innovative solutions.

The following usage cases have been identified as the most interesting for the WING/WORLD toolkit.

i. *Increased coverage in outdoor WiMAX networks*. A hybrid WiMAX/WiFi mesh architecture can decrease the number of WiMAX base stations needed to obtain good coverage. An umbrella coverage model can be envisioned where WiMax technology is used at the third tier of the network architecture to provide connectivity to a multitude of WiFi-based WMNs. Albeit only Ethernet, WiFi, and UMTS backhauls are currently supported by our platforms, its modular design can be easily extended to support other technologies using a dedicated API.

ii. *Reduced dead zones*. In both broadband home networking and enterprise networks, wireless coverage is typically realized using IEEE 802.11 WLANs. In such a scenario, the location of the access points in such a way to avoid dead zones is a major issue, while deploying multiple hotspots is also problematic due to both the costs involved and the necessity of interconnecting each access point using a wired backhaul. In this scenario, the flexibility of the WING/WORLD toolkit allows incremental and economical deployment of indoor WMNs using off-the-shelf wireless router preloaded with the WING/WORLD toolkit. Moreover, the embedded CPE imposes no additional hardware/software requirements on the client-side, making it possible to leverage the entire WiFi installed base.

iii. *Fast provisioning.* The self-configuring mesh backhaul allows for drop-in network coverage only when/where needed. In case of a fully outdoor deployment, no additional equipment needs to be installed on the end-user side. WING is an excellent plug-in solution in existing and less reliable WiFi deployments. In this scenario, the embedded AAA features allow straightforward management of the user base.

7. DISCUSSION

In this paper, we presented WING/WORLD, an open wireless mesh networking toolkit developed with the purpose of enabling realistic studies and performance evaluation of novel technologies and protocols. The testbed is based on off-the-shelf hardware components (i.e., IEEE 802.11 commodity devices) and open software modules. It is based on an "open" philosophy and aims at providing the scientific community with a suitable tool for performance analysis of WMNs that can be easily replicated in research centers around the globe.

WING/WORLD differs from already existing prototypes in that it provides a practical platform that can be used to implement high-capacity WMNs solutions that can be used by the research community to design and test innovative solutions in a realistic environment as well as by end-users and WISPs to deploy community and metropolitan access networks. In order to achieve such goals, we designed and developed a flexible wireless mesh networking toolkit supporting the following.

i. A self-configuring and self-healing modular backhaul technology enabling mesh nodes to automatically detect if they are relays or gateways. The node auto-configures itself as a gateway if an IP address can be obtained using DHCP over one of its gateway technologies (Ethernet, WiFi, and UMTS).

ii. An architecture for achieving both service differentiation and performance isolation (at layer 2.5) in IEEE 802.11-based WMNs. While not providing strict QoS performance bounds, the proposed scheme aims at enhancing the perceived quality of experience by combining opportunistic scheduling and packet aggregation in IEEE 802.11-based WMNs and by implementing a DiffServ-like architecture in order to provide traffic prioritization.

iii. A multiradio/multichannel wireless backhaul capable of finding high performances routing paths in an environment characterized by competing traffic flows that cause link losses, as well as significant link fluctuations.

iv. A Captive portal based end-users management system based on Coovachilli (a well-established access control tool) and exploiting RADIUS as AAA back-end. Such a solution allowed us to open the testbed to both students and staff in our lab, which are using the WMN for their daily Internet tasks.

Current efforts are devoted at devising a unified network monitoring and management architecture capable of supporting the testbed operation across its whole life cycle starting from network deployment and profiling, to software configuration and fault recovery. The challenges in this area are to automate fault-management in WMN and consequently enable the rapid deployment of WMNs. Some solutions are already available from commercial vendors; however, the distributed and decentralized nature of the wireless mesh networking paradigm has a serious impact on the design of robust and performing architectures. A viable architecture must also support real-time node status monitoring, experiment campaign planning and execution, and data gathering/analysis facilities. Finally, a user-friendly interface must be provided in order to enable effective network management. Finally, we are currently extending the software toolkit to support multiple radio interface using a joint routing and channel assignment scheme derived from the ITACA algorithm and exploiting a novel channel aware routing metric. Preliminary results show a significant increase in the aggregated network throughput due to the better spatial reuse of the wireless medium.

Future work is aimed at further improving the self-adaptation capabilities of the WING/WORLD toolkit by borrowing concept from cognitive radios and cognitive networks with the purpose of enabling the validation of several interesting concepts nowadays present in the scientific literature but still in the design/analysis phase.

ACKNOWLEDGMENT

This work was partially supported by the Italian Ministry of Scientific Research (MIUR) within the framework of the Wireless Mesh Network for Next-Generation Internet (WING) project (Grant number RBIN04M292) and of the Wireless multiplatform mimo active access networks for QoS-demanding muLtimedia Delivery (WORLD) project (Grant number 2007R989S).

AUTHORS' AFFILIATIONS

(1) DISI—University of Trento

(2) CREATE-NET

REFERENCES

1. *Akyildiz IF, Wang X, Wang W: Wireless mesh networks: a survey. Computer Networks 2005, 47(4):445-487.*
2. *Bruno R, Conti M, Gregori E: Mesh networks: commodity multihop ad hoc networks. IEEE Communications Magazine 2005, 43(3):123-131.*
3. *Tropos http://www.tropos.com/*
4. *Firetide http://www.firetide.com/*
5. *Strix http://www.strix.com/*
6. *LocustWorld http://www.locustworld.com/*
7. *Meshdynamics http://www.meshdynamics.com/*
8. *Wing Project http://www.wing.project.org*
9. *World Project http://www.world.project.olterviste.org*
10. *Chlamtac I, Conti M, Liu JJ-N: Mobile ad hoc networking: imperatives and challenges. Ad Hoc Networks 2003, 1(1):13-64. 10.1016/S1570-8705(03)00013-1*
11. *Draves R, Padhye J, Zill B: Comparison of routing metrics for static multi-hop wireless networks. Proceedings of the ACM Conference on Computer Communications (SIGCOMM '04), 2004, Portland, Ore, USA 133-144.*
12. *Das SM, Pucha H, Koutsonikolas D, Hu YC, Peroulis D: DMesh: incorporating practical directional antennas in multichannel wireless mesh networks. IEEE Journal on Selected Areas in Communications 2006, 24(11):2028-2039.*
13. *Chowdhury KR, Akyildiz IF: Cognitive wireless mesh networks with dynamic spectrum access. IEEE Journal on Selected Areas in Communications 2008, 26(1):168-181.*

14. *Ghamri-Doudane S, Parker D, Agoulmine N: A cluster-based middleware for infrastructure wireless Mesh networks. Proceedings of the IEEE Wireless Communications and Networking Conference (WCNC '08), 2008, Las Vegas, Nev, USA 3045-3050.*
15. *Riggio R, Miorandi D, Chlamtac I, et al.: Hardware and software solutions for wireless mesh network testbeds. IEEE Communications Magazine 2008, 46(6):156-162.*
16. *OpenWRT Linux Distribution http://openwrt.org/*
17. *GNU General Public License http://www.gnu.org/copyleft/gpl.html*
18. *Case J, Fedor M, Schoffstall M, Davin J: A simple network management protocol. IETF RFC 1157, May 1990, http://www.ietf.org/rfc/rfc1157.txtView Article*
19. *Rosenberg J, Mahy R, Matthews P, Wing D: Session Traversal Utilities for NAT (STUN). IETF RFC 5389, October 2008, http://www.ietf.org/rfc/rfc5389.txtView Article*
20. *Kohler E, Morris R, Chen B, Jannotti J, Kaashoek MF: The click modular router. ACM Transactions on Computer Systems 2000, 18(3):263-297. 10.1145/354871.354874*
21. *Bicket A, Aguayo D, Biswas S, Morris R: Architecture and evaluation of an unplanned 802.11b mesh network. Proceedings of the Annual International Conference on Mobile Computing and Networking (MOBICOM '05), 2005, Cologne, Germany 31-42*
22. *Draves R, Padhye J, Zill B: Routing in multi-radio, multi-hop wireless mesh networks. Proceedings of the Annual International Conference on Mobile Computing and Networking (MOBICOM '04), 2004, Philadelphia, Pa, USA 114-128.*
23. *Riggio R, Miorandi D, Chlamtac I: Airtime deficit round robin (ADRR) packet scheduling algorithm. Proceedings of the 5th IEEE International Conference on Mobile Ad-Hoc and Sensor Systems (MASS '08), October 2008, Atlanta, Ga, USA 647-652.*
24. *Riggio R, De Pellegrini F, Scalabrino N, Li P, Fang Y, Chlamtac I: Performance of a novel adaptive traffic aggregation scheme for wireless mesh networks. Proceedings of the IEEE Military Communications Conference (MILCOM '07), 2007, Orlando, Fla, USA*
25. *Riggio R, Miorandi D, De Pellegrini F, Granelli F, Chlamtac I: A traffic aggregation and differentiation scheme for enhanced QoS in IEEE 802.11-based wireless mesh networks. Computer Communications 2008, 31(7):1290-1300. 10.1016/j.comcom.2008.01.037*

26. *Blake S, Black D, Carlson M, Davies E, Wang Z, Weiss W: An architecture for differentiated services. IETF RFC 2475, December 1998, http://www.ietf.org/rfc/rfc2475.txtView Article*
27. *The olsr.org OLSR daemon http://www.olsr.org/*
28. *B.A.T.M.A.N., Better Approach to Mobile Ad-Hoc Networking http://www.open-mesh.net/*
29. *Coding of speech at 8 kbit/s using conjugate-structure algebraic-code-excited linear prediction (CS-ACELP) ITU-T Recommendation G.729, June 2007*
30. *Slongo M, Rasheed T: ITACA: channel assignment in wireless mesh networks. CREATE-NET; 2008.http://wing-project.org/*
31. *ns-2 http://www.isi.edu/nsnam/ns*
32. *Ramachandran KN, Belding EM, Almeroth KC, Buddhikot MM: Interference-aware channel assignment in multi-radio wireless mesh networks. Proceedings of the 25th IEEE International Conference on Computer Communications (INFOCOM '06), 2006, Barcellona, Spain*
33. *Heusse M, Rousseau F, Berger-Sabbatel G, Duda A: Performance anomaly of 802.11b. Proceedings of the IEEE International Conference on Computer Communications (INFOCOM '03), 2003, San Francisco, Calif, USA 2: 836-843.*
34. *Tse DNC, Hanly SV: Multiaccess fading channels—part I: polymatroid structure, optimal resource allocation and throughput capacities. IEEE Transactions on Information Theory 1998, 44(7):2796-2815. 10.1109/18.737513*
35. *Ganguly S, Navda V, Kim K, et al.: Performance optimizations for deploying VoIP services in mesh networks. IEEE Journal on Selected Areas in Communications 2006, 24(11):2147-2158.*
36. *Willens S, Rubens A, Simpson W: Remote Authentication Dial in User Service (RADIUS). IETF RFC 2865, June 2008, http://www.ietf.org/rfc/rfc2865.txt*
37. *CoovaChilli Access Controller http://www.coova.org/CoovaChilli*
38. *The FreeRADIUS Project http://freeradius.org/*
39. *Johnson D, Hu Y, Maltz D: The dynamic source routing protocol (DSR) for mobile ad hoc networks for IPv4. IETF RFC 4728, February 2007, http://www.ietf.org/rfc/rfc4728.txtView Article*
40. *Ko B-J, Misra V, Padhye J, Rubenstein D: Distributed channel assignment in multi-radio 802.11 mesh networks. Proceedings of the IEEE Wireless*

Communications and Networking Conference (WCNC '07), 2007, Hong Kong 3981-3986.

41. *Ramachandran K, Sheriff I, Belding EM, Almeroth K: A multi-radio 802.11 mesh network architecture. Mobile Networks and Applications 2008, 13(1-2):132-146. 10.1007/s11036-008-0026-8*
42. *Purdue university wireless mesh network testbed https://engineering.purdue.edu/mesh/*
43. *Koksal CE, Balakrishnan H: Quality-aware routing metrics for time-varying wireless mesh networks. IEEE Journal on Selected Areas in Communications 2006, 24(11):1984-1994.*

10

CHAPTER

ACCELEROMETER-ASSISTED 802.11 RATE ADAPTATION ON MOBILE WIFI ACCESS

Yu-Jen Lai[1], Wei-Hao Kuo[1], Wan-Ting Chiu[1] and Hung-Yu Wei[2*]

ABSTRACT

The expansion of 802.11 AP deployment provides opportunistic Wi-Fi access in underground mass rapid transit (MRT) system. However, such vehicular network faces the challenge of limited time for the MS on the train to connect the base station at the station. Therefore, to maximize the throughput within these several tens of seconds becomes crucial to the network. To achieve this goal, we propose accelerometer-assisted rate adaptation (AARA), a mechanism utilizes the out-of-band information of a train's acceleration to improve the conventional rate adaptation scheme. AARA consists of two parts: First, AARA divides a train's movement into four phases and performs real-time estimation on the train's current movement phase. Second, AARA employs the estimation results to enhance the bit-rate selection during each phase. We conduct experiments on two different Taipei MRT systems: high-capacity MRT and medium-capacity MRT. The experimental results show that the average throughput of AARA outperforms that of the conventional scheme in different scenarios. In addition, we also design a mechanism of power saving with the aid of the movement estimation.

Keywords Rate adaptation IEEE 802.11 Sensor-assisted communications Vehicular network Network experiment Public transportation system

INTRODUCTION

As the public transportation system rapidly develops, demands to access the Internet in transportation systems are increasing over time. Immediate information access to the Internet is especially important and valuable for transportation system commuters in metropolises. Due to the development in wireless technology and the expansion of WLAN deployment, this type of vehicular network becomes attainable nowadays. Consider the scenario of Figure1, base station (BS) is located at a roadside spot like a train station, a bus stop, or an intersection to provide Wi-Fi service to mobile station (MS) on vehicles. However, due to limited transmission coverage, wireless signal disappears after the vehicle departs from the station. Therefore, selecting appropriate transmission rates to employ is vital to maximize the transmission throughput when the train stops at the station.

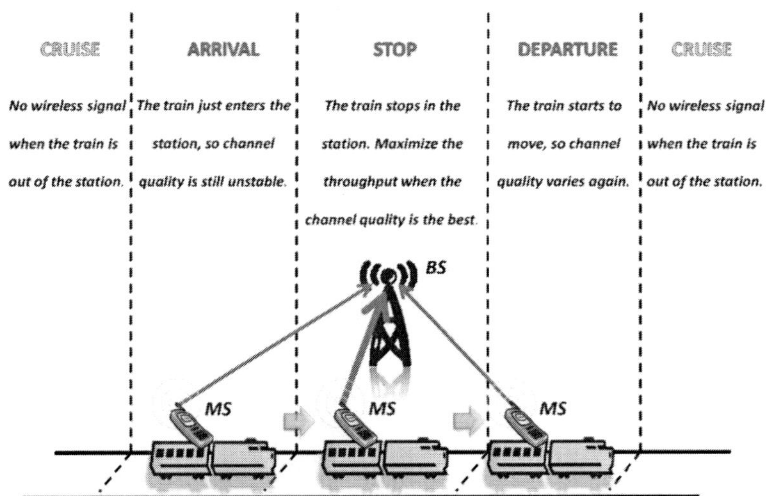

Figure 1. THE SCENARIO OF MOBILE WIFI ACCESS.

In the mass rapid transit (MRT) system, the channel quality varies greatly when the train arrives at and departs from the station, but is relatively stable when the train stops at the station. Most of the conventional rate adaptation schemes suffer from the degradation of transmission performance since the signal environment changes radically when trains passing by the station. To solve this problem, it is conceivable to employ different rate adaptation schemes according to the phases of the train's movement.

In addition, since Wi-Fi has fairly high power consumption when the NIC is activated, we should take energy consumption in mobile devices into consideration. As the train leaves the station, the MS on the train is outside of the Wi-Fi coverage. If we can detect the train's departure, we will be able to switch the wireless device to sleep mode in order to conserve energy. Moreover, we should also recognize when the train arrives at the next station and reactivate the wireless device to relish Internet service again.

Therefore, the first challenge is to accurately estimate the movement of the train. Intuitively, we would attempt to employ the metric of signal strength to estimate the train's distance from the station. Nevertheless, this method requires extra energy to monitor the wireless channel along the whole path, even when it is out of signal coverage, i.e., the train is cruising between stations. Thus, in this article, we have designed a context-aware mechanism assisted by out-of-band acceleration information to estimate the movement of the train. We also propose a rate adaptation scheme assisted by the estimation result of the train's movement in mobile WiFi devices. The experimental results demonstrated the feasibility of the proposed scheme. In our current prototyping, the accelerometer is attached to the mobile device. As acceleration information has been shown useful to assist wireless communications, we could seek to obtain acceleration information with various methods. For example, for radio system built-in for train vehicles, the acceleration information might possibly obtained from train control system. For radio system on motorcycles, the acceleration information might be obtained from its control panel. Accelerometer could be built-in internally (e.g. smartphones have built-in accelerometer) or connected externally (e.g. connected via USB port). We foresee this acceleration-assisted wireless communication technology to be applied in different forms.

The rest of this article is organized as follows. Section "Related work" discusses the related work. Section "Design overview" gives an overview of the system design. Section "Preliminary experiments" presents the preliminary data collection and observation from the preliminary experiments. Section "System design" introduces the detailed design of accelerometer-assisted rate adaptation (AARA). Section "Experimental evaluation" presents the comparison of the evaluation results of AARA and SampleRate on Taipei MRT. Section "Power saving" compares the energy consumptions of AARA and SampleRate. Section "Discussion" discusses the applicability of the proposed scheme. Section "Conclusion" concludes the article.

RELATED WORK

Movement estimation

GPS has been utilized to estimate the moving speed or people's behavior. Zheng et al.[1] analyze GPS logs with graph-based post-processing algorithm to identify transportation modes. Patterson et al.[2] combine GPS with a learned Bayesian model to infer the transportation modes. Reddy et al.[3] solve the same issue by dealing with GPS and acceleration information based on a discrete Hidden Markov Model. However, instead of estimating the vehicle's departure and arrival, these articles tend to focus on recognizing human behaviors. More than that, using the information of GPS location requires the user to keep the records of the BS's GPS location in advance, and the GPS signal cannot be received in underground scenarios.

Rate adaptation

Typically, packet transmission's statistics of success and failure are the indicators of channel conditions, which are used by wireless LAN devices to adjust the bit-rate. Auto rate fallback (ARF) is the first rate adaptation scheme designed for WaveLan II devices. Lacage et al. design adaptive auto rate fallback (AARF)[4] to further improve ARF. After a fixed number of consecutive successful transmissions, both ARF and AARF increase their bit-rate and then decrease it after some transmission failures occur. The main difference between them is that AARF doubles the number of consecutive successful transmissions needed to switch to a higher bit-rate when transmission failure occurs. Onoe[5] is a credit-based algorithm and has been implemented in Linux MadWifi driver. It computes the credit of each bit-rate per second to decide whether to increase, decrease, or remain the bit-rate. When the retransmission times are higher than a certain level, it will then decrease the credit of the given bit-rate. RRAA[6] monitors the short term loss ratio (during a window) in a given bit-rate and compares the loss ratio with two pre-defined thresholds called P_{ORI} and P_{MTL}. When the loss ratio exceeds P_{MTL}, the bit-rate decreases; while the loss ratio is below P_{ORI}, the bit-rate increases. The idea of RRAA is to switch to a bit-rate only when it can supply a higher throughput than the original bit-rate.

SampleRate is the default rate adaptation algorithm in MadWifi driver[7]. The idea of SampleRate is to choose the bit-rate scheme with the shortest expected transmission time to transmit. To achieve this, it maintains a table of the average successful transmission time of each bit-rate, and sends probing packets periodically to update the table. For probing efficiency, SampleRate blocks (not to probe) the bit-rates which have consecutive transmission failures.

CONTEXT-AWARE WIRELESS TRANSMISSION

Most of the current rate adaptation mechanisms are designed for wireless networks in general scenarios. Different from the previous rate adaptation schemes, the proposed rate adaptation mechanism is aware of the context assisted by out-of-band acceleration information designed for wireless access on mobile environment. CARS[8] is focused on rate adaptation for vehicular networks. With GPS devices, CARS obtains context information like vehicle speed and distance to the receiver. Rahmati and Zhong[9] also consider employing context information to optimize the energy by switching wireless interface between Wi-Fi and GSM. Joshi and Kulkarni's[10] work compares the rate adaptation schemes implemented in MadWifi, and address a rate adaptation scheme based on history record of the relation between SNR and throughput. Santhapuri et al.[11] discuss the application of out-of-band context information from sensors to assist wireless communication design. Note that we choose context information from accelerometers rather than GPS locations, as it incurs several problems while employing information of GPS locations. (1) Underground metro system fails to access GPS signal. (2) Problems like weak signal and long GPS acquisition/re-acquisition time occur in urban settings with tall building. (3) Prior knowledge of the access point locations is needed. (4) GPS consumes more than ten times of power than that of accelerometer.

DESIGN OVERVIEW

This section describes our system's design overview. The main contribution we are making is to boost the throughput with the aid of a more suitable rate adaptation strategy for the scenario of "drive-thru with a stop", as shown in Figure1.

The drive-thru scenario, brought up by Ott and Kutscher[12], refers to a scenario in which the vehicle drives through a roadside unit and access it in a very short time. Many related works adopt different methods to increase the throughput in such scenario[13–17]. On the basis of this concept, we observe that the scenario of drive-thru with a stop should be even more critical as it supplies a higher connection opportunity when the vehicle stops near a roadside unit. For example, if there is a municipal plan to deploy Wi-Fi AP on a specific road for building a vehicular network, it would be more reasonable to deploy APs at an intersection, where most vehicles might stop for a while. Similarly, the most sensible place to deploy APs in the public transportation system is in the station, where the train also stops for a short period of time.

Even so, the connection time is still limited. In our testbed, the Taipei MRT system, it takes about 40–50 s for a train to approach to, stop at, and depart from a station. Due to the changing channel quality along the path, we need a more suitable rate adaptation scheme specifically designed for the drive-thru with a stop scenario.

In this article, we propose a novel rate adaptation scheme, AARA, assisted by accelerometer that mainly designed for transportation systems. As illustrated in Figure2, the mechanism of AARA is composed of two parts: the estimation of movement phases and the scheme for rate adaptation.

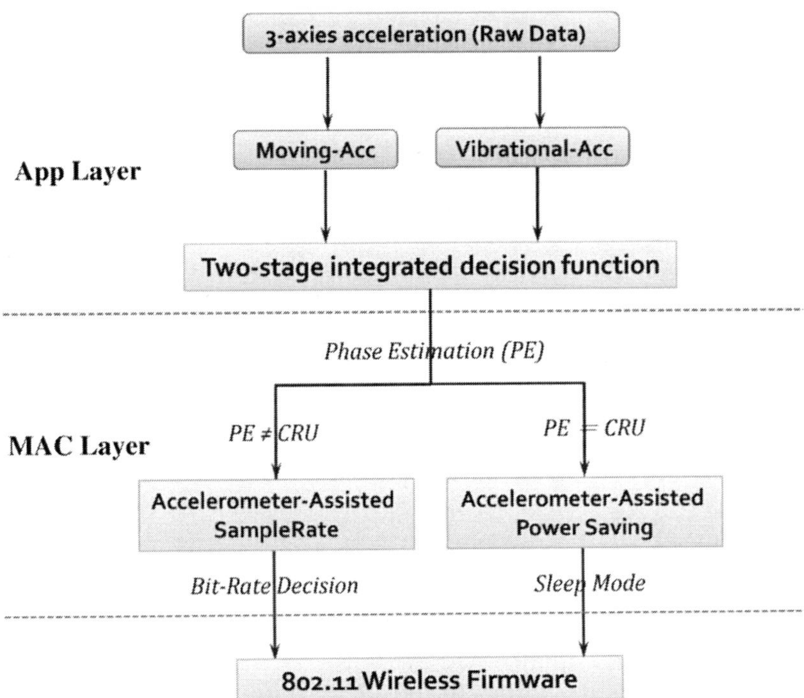

Figure 2. THE SYSTEM ARCHITECTURE OF AARA.

Accelerometer-assisted rate adaptation measures the 3-axis acceleration of the MS, and transforms the data into two indicators: Moving-Acc (M-Acc) and Vibrational-Acc (V-Acc). These indicators describe the acceleration of a train's movement and vibration, respectively. The phase estimation is acquired from a simple integrated decision function. The system passes the estimation result to the MAC layer to facilitate the bit-rate selection as well as the power saving function. The proposed rate adaptation scheme, assisted by accelerometer, is an enhancement to the SampleRate algorithm, which is the default bit-rate selection mechanism in MadWifi driver. Moreover, the phase estimation is achieved in real-time and the system is implemented and tested in the real-world experiments.

PRELIMINARY EXPERIMENTS

Experiment setup

Before introducing our system design in detail, we summarize the characteristics of our experimental testbed, the equipments, the platform, and the methodology in this section. The experimental results presented in this article are obtained by conducting real-world experiments in outdoor public transportation system (Taipei MRT).

Methodology

We conduct the experiments with three laptops operating on Linux. We choose MadWifi[18] as the driver of the wireless adaptor since it is one of the most advanced open-source WLAN driver available for Linux. The detailed parameters of our experiments are presented in Table1.

The scenario of our system is shown in Figure1. We place two laptops on the train and one on the platform as the access point. One of the laptops on the train acted as an Iperf client[19], taking charge of continuous transmission to the access point situated on the platform which acted as Iperf server. The other laptop on the train, on the other hand, acted as a sniffer running in the monitor mode and using Wireshark[20] to sniff and log the information of the transmission packets. The information of packets' bit rates are extracted from the Radiotap header[21]. Meanwhile, the client on the train also runs the accelerometer, attached via USB, to measure its 3-axis acceleration. The raw data of 3-axis acceleration is collected and processed into two types of information to estimate the train's movement phases. Next, we use the estimation result to improve the transmission strategy accordingly. We further describe our design in Section "System design".

Table 1. Experimental parameters

Parameters	Values
Computer	Toshiba Satellite M200
Operating system	Linux Kernel 2.6.28-18
Wireless adaptor	DWL-AG660
Wireless driver	MadWifi v0.9.4
Protocol	802.11 b/g
Frequency	2.4 GHz
Transmission power	63 mW
Traffic generator	Iperf - UDP
Payload size	1470 Mbytes
CBR rate	54 Mbps
Accelerometer	Hitachi H48C

Transportation system characteristics

Since our system aims to be widely workable in different train stations and transportation systems, we conduct our experiments on both the High-Capacity MRT and the Medium-Capacity MRT, each with its own movement patterns. There are two types of MRT systems in Taipei MRT, the High-Capacity MRT and Medium-Capacity MRT[22]. These two systems have different rolling stocks, control modes (automatically driven or driver driven), platforms, tracks, and wheels. The time for the train to speed up and to slow down in the station are also different in the two MRT systems. For example, the time the train takes from entering the station to stopping at the platform is about 12 s in Medium-Capacity MRT, and 18 s in High-Capacity MRT, as shown in Figure3. This difference is due to the platform lengths and the numbers of car in two kinds of MRT systems. To understand more about the moving patterns in these two MRT systems, we conduct experiments to collect data in advance. For simplicity, we abbreviate High-Capacity MRT and Medium-Capacity MRT as HC-MRT and MC-MRT, respectively, in the rest of this article.

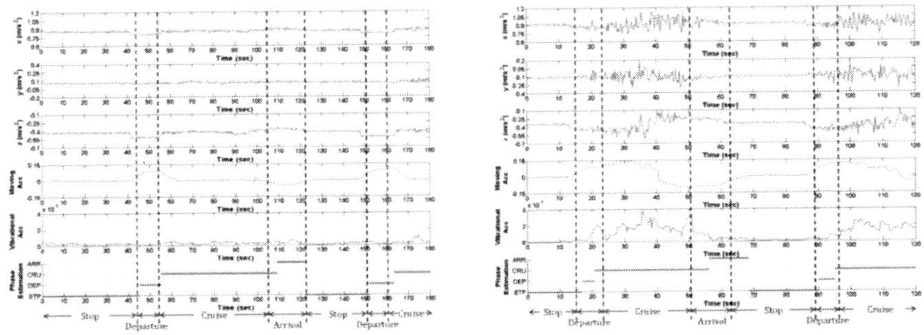

(a) The phases estimation in High-Capacity MRT (b) The phases estimation in Medium-Capacity MRT

Figure 3. THE RAW DATA OF **x,y,z** AXIS ACCELERATION, V-ACC, M-ACC, AND THE ESTIMATED PHASES IN EITHER TYPE OF MRT SYSTEMS. The dotted lines show the partitions of the actual movement phase.

We conduct our experiments on Wenshan-Neihu Line for MC-MRT, on Xiaonanmen Line and Nangang Line for HC-MRT, as shown in Figure4. We carry the client and sniffer nodes on the train moving from Taipei Zoo station (start) to Xinhai station (end), and the server node situated in Wanfang Community station (middle) is in between. When the train arrives at Wanfang Community station (middle), the node starts to transmit and lasts for a while until the train departs from the station. The experiments conducted in HC-MRT have similar experimental settings.

(a) Xiaonanmen Line (High-Capacity System) (b) WenHu Line (Medium-Capacity System)

Figure 4. THE EXPERIMENTAL TOPOLOGIES OF THE TWO MRT SYSTEMS.

Preliminary results

Prior to estimating the movement of the train with accelerometers, we must perform preliminary experiments to investigate the acceleration properties of MRTs. We conduct the following experiments: tilting the accelerometer in different angles, setting the client at the side or the middle of the train, collecting raw data in different stations, and perform the experiments in different types of MRT systems. After analyzing the experimental data, we conclude that the most critical factors are the orientation of the accelerometer and the MRT system type.

In this article, we categorize the train's movement into four phases according to its behavior: Stop (STP), Departure (DEP), Cruise (CRU), and Arrival (ARR). STP phase refers to the duration the train stops at the station. The DEP phase is the time the train starts to move and then accelerates. In CRU phase, the train moves between the stations at a relatively steady speed. When the train approaches to the station and slows down, it enters the ARR phase. This also represents the process a passenger gets on the train, stays on the train, and gets off the train. This process has a total of five phases:

$$\text{STP(get on)} \rightarrow \text{DEP} \rightarrow \text{CRU} \rightarrow \text{ARR} \rightarrow \text{STP(get off)}$$

Figure5 shows the raw data of 3-axis acceleration when the train passes through a station in MC-MRT and HC-MRT, respectively. Note that our goal is to estimate a train's movement regardless of the position and orientation of users' accelerometers. Thus, we collect the information of acceleration when the accelerometer is set to head-orientated and tilted orientation. In this article, head-orientated refers to when the x-direction of the accelerometer points towards the first car, the y-direction towards the right side of the train when the user faces the first car, and the z-direction towards the floor. The tilted orientation refers to when the accelerometer is randomly placed. For instance, we put the accelerometer in our bags, or hold it at different orientations.

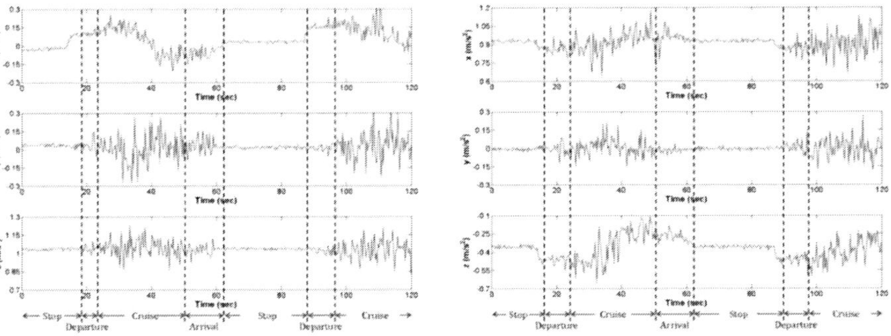

(a) Acceleration in Medium-Capacity MRT (Head-orientated)

(b) Acceleration in Medium-Capacity MRT (Tilted)

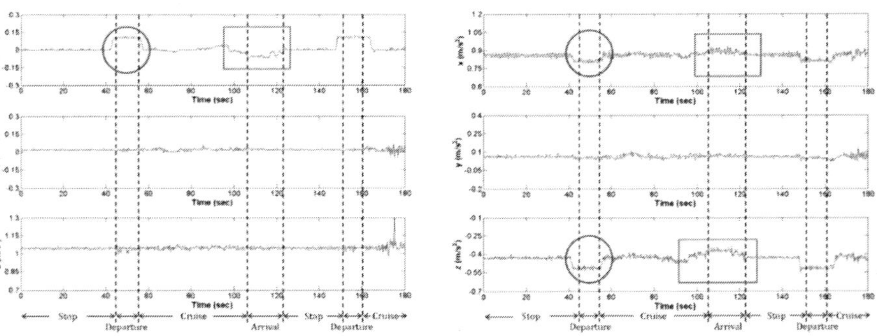

(c) Acceleration in High-Capacity MRT (Head-orientated)

(d) Acceleration in High-Capacity MRT (Tilted)

Figure 5. THE RAW DATA OF x,y,z AXIS ACCELERATION IN HC-MRT AND MC-MRT SYSTEMS. The dotted lines divide the actual four movement phases.

As shown in Figure 5a,b, we observe that the vibration of acceleration is more acute in MC-MRT. The vibration level is lowest in STP phase, and when the train starts to move (DEP phase), the vibration level immediately elevates. The amplitude of vibration reaches its maximum in CRU phase, and decreases again when the train approaches to the station (ARR phase). The vibration of acceleration turns stable when the train fully stops at the station. The experimental results in both the head-orientated scenario and the tilted scenario are shown to be similar in the MC-MRT.

On the other hand, in Figure 5c,d, the vibration level in HC-MRT is lower than it is in MC-MRT. However, in Figure 5c, we notice that there is a rise in acceleration in the *x*-axis during DEP phase (the red circle) and a decline when the train approaches to the station (the green rectangle). These rise and decline are actually the acceleration generated from the train's speeding up and slowing down. We can see the acceleration in *x*-axis increases when the train starts to move, and decreases to about 0 when the train enters CRU phase, which means the train is moving with a steady speed. When the train approaches to the station, the acceleration in the *x*-axis drops again in the negative region (starting from around 100 s) because the train's direction of acceleration is now in reverse. Finally, the acceleration is back to around 0 when the train fully stops in the station. Though the acceleration pattern in ARR phase is not as obvious as that in DEP phase, it is still feasible to exploit the information to estimate the movement phases.

Figure 5d is the measured result of acceleration in HC-MRT system when the accelerometer is tilted. As shown in the circled areas of Figure 5d, although the acceleration patterns are less noticeable, they can still be observed from *x*-axis and *z*-axis. The main reason is that acceleration information scatters into two or three axes. This results in a mitigation of the signal intensity. We can also observe that the "rise" during DEP phase in Figure 5c becomes a "decline" in Figure 5d because we have arbitrarily set the orientation of accelerometer. The direction of *x*-axis of the accelerometer might be opposite to the train's moving direction. These observations infer that we can not estimate the phases solely by relying on information from one specific axis.

From the observation above, we conceive that the information from the acceleration should be classified into two parts: vibration and movement. These information reside in the high frequency and low frequency of the acceleration, respectively. In the next section, we illustrate the design of our estimation scheme and how it extracts information from both parts to enhance the estimation accuracy.

SYSTEM DESIGN

In this section, we describe the system design and the implementation of AARA. The details of the two components, movement phase estimation and rate adaptation scheme, are respectively introduced in the following two subsections.

Movement phase estimation by accelerometer

Our goal is to estimate in real-time which movement phase the train is in: STP, DEP, CRU, or ARR. The estimation mechanism has the following properties:

- robust against the accelerometer's orientation and position
- previous off-line training not required
- user's knowledge about the environments not required

The proposed scheme designed for both HC-MRT and MC-MRT excluded the need of user's knowledge about the environments. In other words, the users neither need to input the information about which type of MRT system they are taking nor make the accelerometer's orientation pointing to a specific direction. However, as described previously, we notice that different transportation systems have different moving patterns. It is hard to estimate the train's movement simply based on the raw data or any single feature. Other than that, as shown in Section"Preliminary results", the measured acceleration is composed of two parts: high frequency and low frequency. The high frequency indicates the short-term vibration of the train, whereas the low frequency shows its long-term movement.

In HC-MRT system, the train's vibration level is relatively low, but the train's speeding up and slowing down patterns are more obvious. By contrast, in MC-MRT system, the train's vibration level is high, but the train's speeding up and slowing down patterns are less obvious. In order to estimate the movement phases based on these two characteristics that apply to different MRT systems, we define those two indicators of acceleration as M-Acc and V-Acc. M-Acc depicts the train's movement, and V-Acc represents the vibration of the train.

The two indicators are defined as follows:

$$\text{Vibrational}-\text{Acc} = \text{Var}(||A||-||G||),$$

$$\text{Moving}-\text{Acc} = ||\text{LP}-\text{Filter}(\vec{M}_{\text{est}})|| \times \text{sgn}(\vec{M}_{\text{est}} \cdot \vec{M}_{\text{DEP}}),$$

Where $\hat{A}$ is the original measurement of the 3-axis accelerations,$G^{\rightarrow}$ is the gravitation, $\vec{M}_{\text{est}} = \hat{A} - \vec{G} - \vec{D}_{\text{off}}$ is the estimated moving acceleration of the train,$D^{\rightarrow}$ off D→off is the bias offset due to the imperfection of the accelerometer, and$M^{\rightarrow}$ DEP M→DEP is the estimated moving acceleration recorded in the DEP phase for directional reference.

V-Acc & M-Acc

We define V-Acc as an indicator of the vibration level, and thus we simply compute the variance of the measure's length minus the gravitation's length. The length of an vector is calculated $\sqrt{x^2 + y^2 + z^2}$. Note that we set $||\vec{G}||$ as 1 since the value of gravitation is typically 1×g. The size of a sliding window for computing the variance of the difference between the measurement and gravitation is set to be 20 data points. This means we calculate 20 sets of $(||\hat{A}|| - ||\vec{G}||)$, and then calculate their variance. Since the sampling rate of accelerometer is about 0.25 s, the latency of estimation is tolerable. In addition, the size of sliding window is also feasible to estimate the acceleration of a train's vibration.

As for M-Acc, we use Use Use Figure6 to illustrate the relationship between the acceleration components. We denote the acceleration of a train's speeding up and slowing down as $\vec{M}$. The vector $\vec{M}$ varies over time, and is actually the moving acceleration that we want to predict by M-Acc. In STP phase, the acceleration of the train's movement should be zero vector ($\vec{M}=\vec{0}$). Our measurement, $\hat{A}$, however, is not zero vector due to gravitation and the bias offset generated by the imperfection of the accelerometer. We firstly record ($\vec{G}+\vec{D}$ off) and calibrate$\hat{A}$ to be zero vector by eliminating($\vec{G}+\vec{D}$ off), as shown in Figure 6a.

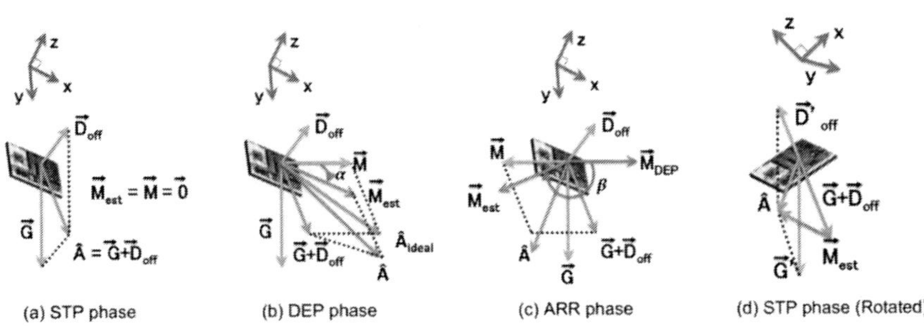

Figure 6. ACCELERATION COMPONENTS OF THE ACCELEROMETER IN DIFFERENT PHASES.

As the train starts to move,$\vec{M}$ increases. In ideal case, the measurement $\hat{A}$ should be $\vec{M}+(\vec{G}+\vec{D}$ off), so that we can extract $\vec{M}$ by eliminating ($\vec{G}+\vec{D}$ off) from . However, the measurement $\hat{A}$ is easily corrupted, as shown in Figure 6b. Thus, after eliminating ($\vec{G}+\vec{D}$ off), the corrupted $\hat{A}$ will be$\vec{M}$ but not $\vec{M}$. We calculate the length of $\vec{M}$ est and apply the

EWMA (exponentially weighted moving average) with $\alpha = 0.075$ as a LP-Filter to prevent the abrupt burst of the measurement. Though it is enough to predict the length of $\vec{M}$, its direction is unknown, so we are not able to distinguish whether the moving acceleration is caused by speeding up or slowing down. Therefore, we record $\vec{M}_{est}$ in DEP phase as $\vec{M}_{DEP}$, which represents the directional reference of the train's movement. And then, we compute the inner-product between $\vec{M}_{est}$ and $\vec{M}_{DEP}$. If the sign of $(\vec{M}_{est} \cdot \vec{M}_{DEP})$ is positive, it means that the angle between $\vec{M}_{est}$ and the direction of train's movement is less than 90° (as shown in Figure 6b, the α angle is less than 90°; thus, the direction of $\vec{M}_{est}$ and the direction of train's movement are in the similar directions).

On the other hand, if the sign of $(\vec{M}_{est} \cdot \vec{M}_{DEP})$ is negative, the angle of the measurement and the direction of the train's movement is greater than 90° (as shown in Figure 6c, the β angle is over 90°, which means the measured $\vec{M}_{est}$ and the direction of train's movement are in opposing directions).

Auto-calibration

As previously stated, we calibrate $\hat{A}$ by removing $(\vec{G} + \vec{D}_{off})$, when the train is in STP phase. We also record $\vec{M}_{est}$ in DEP phase as $\vec{M}_{DEP}$, as a directional reference of the train's movement. AARA performs the auto-calibration in both STP and DEP phases to prevent the propagation of estimation error. In Figure 6d, the orientation of the accelerometer is changed. The vectors of the gravitation and the bias offset in the new vector space are also changed to be $\vec{G'}$ and $\vec{D'}_{off}$, respectively. If we make the measurement $\hat{A}$ minus the same $(\vec{G} + \vec{D}_{off})$ as before, we derive a wrong $\vec{M}_{est}$, as shown in Figure 6d. Therefore, AARA calibrates the $(\vec{G} + \vec{D}_{off})$ in STP phase and the $\vec{M}_{DEP}$ in DEP phase automatically. In other words, the unexpected estimation error will not propagate because the calibration of the estimator will be reset when the train stops at the next station.

Figure 3a,b show the two indicators when a train passes through the stations in either MRT system.

Two-stage integrated decision function

After processing the raw data, the information of movement and vibration are now much clear to estimate the phases of the train's movement. In our design, we utilize a two-stage estimation method to determine the movement phase.

Before introducing the two-stage integrated decision function, we present the behavior of the acceleration in either MRT system. From Figure 3a, we see that the pattern of the M-Acc is neat in HC-MRT system because the train runs more smoothly. In STP phase, the M-Acc is stable. In DEP phase, the M-Acc starts to increase and reaches to the maximum as it changes to CRU phase. In CRU phase, the M-Acc is stable at around 0 most of the time, only with a few sharp fluctuations. In ARR phase, the M-Acc gradually decreases in the negative region and then increases. The result shows the M-Acc has the high correlation with the speed changes of the train in HC-MRT system.

However, in Figure 3b, it is observed that the use of M-Acc alone is incapable to precisely estimate the phases in MC-MRT system. For example, due to the noise caused by the more acute vibration of the car, the M-Acc in the CRU phase is not as stable as that in the HC-MRT system. Hence, these unpredictable shakings of M-Acc in the CRU phase will cause the estimation mechanism misjudge the shakings as the signal of deceleration of the train, and thus misestimate the ARR phase too early. The V-Acc is helpful in this problem. It is clear that the V-Acc is higher when the train is running. We set thresholds of the V-Acc according to it's average value for assisting in the estimation. While the V-Acc is high, the train must be between the stations and not entering the station yet.

The proposed two-stage integrated decision function is able to estimate the movement phases precisely. The first stage is to estimate which type of the train the user is on, the HC-MRT or the MC-MRT, according to the level of V-Acc. Note that the vibration level is very low in STP phase in both of the MRT systems, so we are unable to distinguish which MRT system it is when the passenger just gets on-board. Fortunately, since the pattern of M-Acc from STP phase to DEP phase, and from DEP phase to CRU phase, are very similar in both MRT systems, AARA can determine the time the train enters the DEP and CRU phase, respectively even when AARA does not yet know which type of MRT the passenger is taking. After the train enters CRU phase, the level of V-Acc is considerably different in HC-MRT and MC-MRT, and thus it is able to distinguish the type of MRT system then.

The next stage is to estimate the ARR and STP phases in both MRT systems. The decision function mainly utilize the M-Acc to predict the phases while V-Acc is used to improve estimation accuracy. Since the level of M-Acc and V-Acc are different in the two MRT systems, we set different thresholds to estimate the ARR phase. We present the details and the parameters of the two-stage decision function in Figure7 and Table2. In Table2, T denotes different thresholds in each decision expression, k denotes the multiples of the thresholds, and $t_1 - t_7$ refer to the time duration for the decision expression to stay True so that the decision expression goes to Yes. Take the top decision expression in Figure7 for example, the decision expression shows that it switches from STP phase to DEP phase only when the train is estimated to stay in STP over t_1 (10 s) and when M-Acc is greater than T_{DEP} for t_2(3 s).

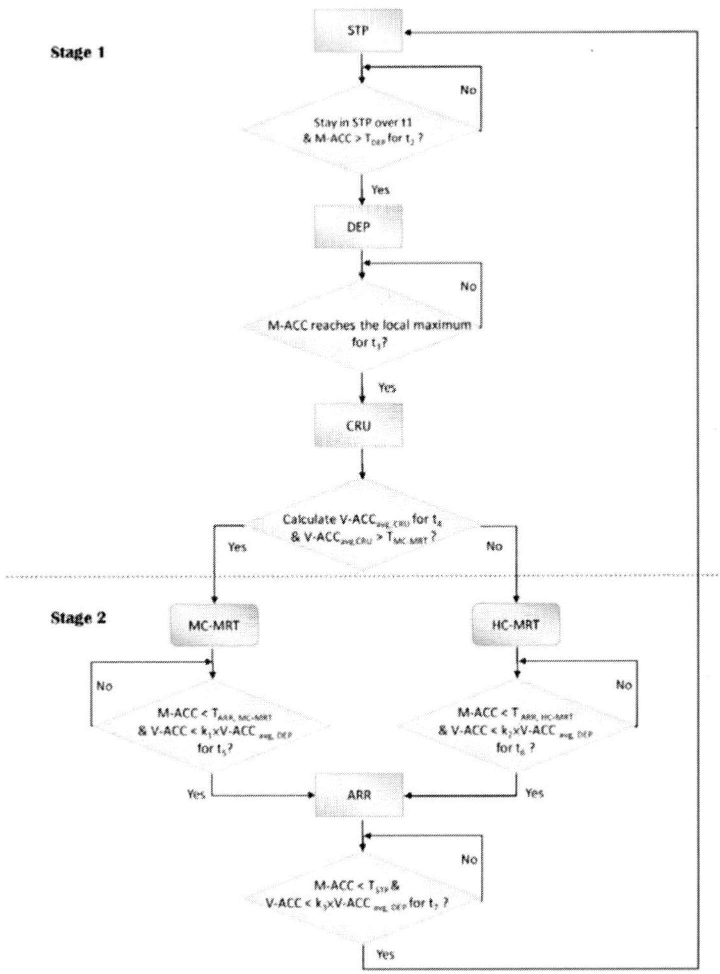

Figure 7. FLOWCHART OF THE TWO-STAGE DECISION FUNCTION.

Table 2. Parameters of the two-stage decision function

Parameters	Values	Parameters	Values (s)
$T_{MC\text{-}MRT}$	0.001	t_1	10
T_{DEP}	0.01	t_2	3
$T_{ARR,\ MC\text{-}MRT}$	−0.02	t_3	1
$T_{ARR,\ HC\text{-}MRT}$	−0.06	t_4	10
T_{STP}	0.02	t_5	2
k_1	3	t_6	2
k_2	2	t_7	1
k_3	3		

We pick two lines from HC-MRT and one line from MC-MRT yielding a total of 11 stations to verify the accuracy of the estimation mechanism. The experimental results are presented in the following section.

Rate adaptation scheme

In this section, we describe how we exploit the phase estimation to enhance the rate adaptation in AARA. We implement AARA by modifying SampleRate bit rate selection algorithm[7]. As described in Section "Related work", SampleRate is the default rate control algorithm of the latest version of MadWifi driver[18]. Its performance has been extensively tested and verified. Since it is the most widely used rate adaptation scheme, we integrate it into our phase estimation to boost the transmission performance.

The concept of our rate adaptation scheme is simple and novel. Since traditional rate adaptation algorithms are designed for general purposes, it has no auxiliary information about the channel quality in general situations. They rely on certain measurements, such as packet error rate or signal strength, to recognize the channel's quality and can only be a passive adaptor to the channel. By contrast, since we know that the channel condition is different when a train passes through a station in each phase, AARA identifies the train's current movement phases to proactively adopt suitable transmission strategies. In other words, the proposed mechanism is a two-tier design. We use the information of the train's movement phases estimated by accelerometer to deal with long-term channel variation, and make use of the rate adaptation scheme, such as SampleRate algorithm, to handle the channel variation in short-term time.

The SampleRate algorithm maintains a table of average successful transmission time of each rate scheme, and it chooses the bit rate scheme with the shortest expected transmission time to transmit, as described in Section "Related work". In ARR phase, when the train is just approaching to the station, there is no history record yet, so SampleRate chooses the highest rate to transmit, e.g. 54 Mbps in 802.11g. However, the channel is poor due to the large distance at this instance, and thus the transmission success rate is low. AARA knows that the channel quality is unstable when the train enters the station, so the MS should adopt a more robust bit rate. Therefore, instead of adopting from the high bit-rates as SampleRate does, AARA adopts from low bit-rates and uses a higher probing rate to adapt to the changing channel. We choose the parameter of probing rate in ARR phase r_0 in Figure8 as 10%, which is the default probing rate in SampleRate algorithm.

AARA Procedure

```
Sense the change of the movement phases
if phase = Arrival then
        Activate the wireless device
        Clear history record
        Adopt from Lowest Bit Rate
        Probing Rate = r0
else if phase = Stop then
        Clean history record
        Adopt from Highest Bit Rate
        Probing Rate = r1
else if phase = Departure then
        Probing Rate = r0
else if phase = Cruise then
        Turn to Sleep mode
end if
```

Figure 8. THE PROCEDURE OF AARA.

When the train stops, the wireless channel turns stable, so the MS should immediately switch to higher rates to maximize the throughput. However, due to the consecutive failure of probing high rate in the previous phase,

SampleRate still considers the channel quality poor, thereby using the lower rate to transmit. AARA clears the stale record and probes from the highest rate as it enters STP phase. Besides, once the MS reaches to its optimal bit-rate, the extra probing packets should be reduced, so the MS can stay using the optimal bit-rate most of the time. Hence, we decrease the probing rate in STP phase (r_1 is set to be 1% in our parameter setting). When the train starts moving, the channel quality degrades promptly as the train speeds up, so the MS should sense the unstable environment and automatically adapt to the channel again. Figure8 illustrates the pseudo code of AARA's procedure.

Experimental evaluation

We test the transmission performance of AARA and SampleRate in Taipei MRT's HC-MRT and MC-MRT systems under the same conditions. The following experimental result shows that AARA can accurately estimate the movement and improve the throughput.

Movement estimation of the trains

In order to adopt different rate adaptation strategies to every movement phase, the most critical issue is to accurately estimate the movement phases. We employ two metrics in the evaluation: recall and delay. Recall is defined as the portion of correct categories that are estimated, and is represented by Estimated-Y/Real-Y. For example, in Figure 3a, there are 67 data points in DEP phase, however, the estimator only estimate 60 data points to be in DEP phase during the real DEP phase; thus, the recall of the estimation of DEP phase is 89.6%. On the other hand, we also measure the estimation delay time, which is a metric more directly related to the user. We define a positive value of delay time to be when the beginning of the estimative phase is later than its actual time. By contrast, a negative delay denotes that the beginning of the estimative phase is earlier than it really is.

The experimental results for the estimative mechanism are summarized in Tables3 and4. We show the recall and delay in each phase in the two types of MRT systems. Note that the result shown in the two tables agrees with the result shown in Figure 3a,b.

Table 3. Results of movement estimation in high-capacity MRT

Phase	Recall (%)	Delay (s)
STP	100	−0.32
DEP	89.6	1.22
CRR	94.5	1.20
ARR	74.6	4.22
Total	94.2	–

Table 4. Results of movement estimation in medium-capacity MRT

Phase	Recall (%)	Delay (s)
STP	92.9	2.98
DEP	55.8	2.04
CRR	100	-2.33
ARR	76.2	3.00
Total	89.4	–

From Tables3 and4, we observe that the total recall in the two MRT systems are 94.2 and 89.4%. The recall is lower in MC-MRT because the train's vibration is severer, causing the M-Acc more disorder than that in HC-MRT. Thus, this effect caused by severer vibration degrades the performance of estimation in MC-MRT. We can see that most of the recall is over 74%, with only one exception. In the DEP phase of MC-MRT, the recall is 55.8%. This is due to some inevitable estimation delay generated by the estimator when using EWMA to smooth M-Acc. Since the duration of DEP phase is relatively short, only 8.37 and 7.19 s in Figure 3b, the effect of delay is more obvious. Actually, from the user's view, there is only 2.04 s delay in estimation of DEP phase in MC-MRT, similar to the delays in other phases.

Rate adaptation in MC-MRT and HC-MRT

In this section, we present the result of the proposed rate adaption scheme assisted by the movement phases estimation. In Figure9, we show the throughput and bit-rate selection of AARA and SampleRate, respectively, during the four movement phases in HC-MRT. The same experiments are also conducted in MC-MRT, as shown in Figure10. We plot the figures of AARA and SampleRate in different figures because the trains' behavior might not totally match in each run. Hence, the duration of the movement phases is also slightly different in the experiments conducted.

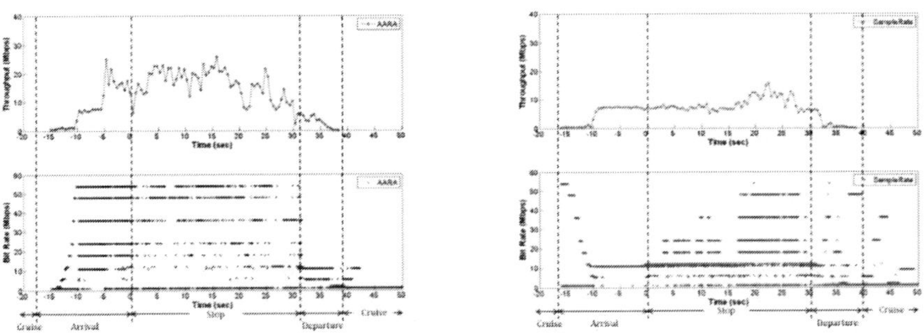

(a) AARA (b) SampleRate

Figure 9. THE THROUGHPUT AND THE RATE SELECTION OF AARA AND SAMPLERATE ALONG THE PATH IN HC-MRT.

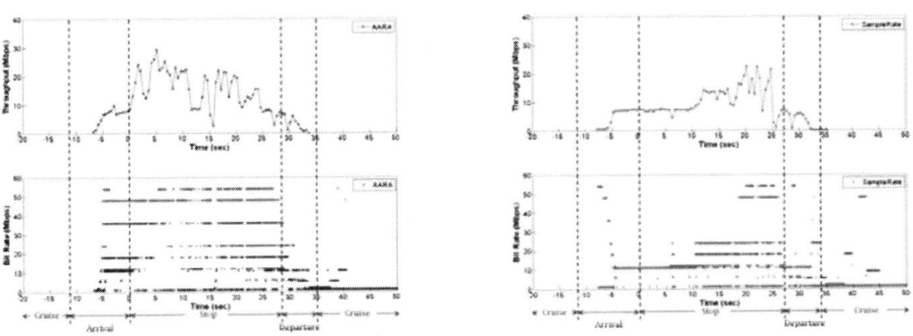

(a) AARA (b) SampleRate

Figure 10. THE THROUGHPUT AND THE RATE SELECTION OF AARA AND SAMPLERATE ALONG THE PATH IN MC-MRT.

First, we observe that AARA achieves significantly higher throughput than that of SampleRate during the four phases. In ARR phase, AARA adopts from the lowest bit-rate because the MS should use a robust bit-rate when the train just enters the station. However, SampleRate chooses the bit-rate with the shortest expected transmission time to adopt, i.e., 54 Mbps in 802.11g. As described in Section "System design", we know that the distance between the MS on the train and the BS in the station is still far, hence, the channel quality is poor. These packets in higher bit-rate mostly fail to transmit. This not only wastes the time to send packets that are unable to be received, but also hinders the bit-rate to increase in the latter part of the ARR phase. In Figure9, we see that the throughput of AARA begins to increase from −5 s, while SampleRate maintains the same throughput level when the train is close to the station (still in ARR phase). This is because the default SampleRate will block the bit-rates with consecutive transmission failures. Since consecutive failures often occur when SampleRate tries the bit-rate from the highest one, it cannot switch to a higher bit-rate even the channel quality is better, i.e., the train is closer to the station but not yet fully stops.

Next, we see the STP phase, the most crucial phase that dominates the overall throughput. In the beginning of STP phase, AARA adjusts to the highest bit-rate because it removes the history transmission statistics, which records the outdated channel condition in the previous phase. On the other hand, SampleRate has no extra information to know that the history record cannot reflect the channel condition anymore. Instead, SampleRate considers that the channel quality is as bad as before; thus, it uses a relatively lower bit-rate, which then downgrades the throughput. SampleRate keeps using lower rates, mostly 11 Mbps, for about 15 s in both Figures 9b and 10b. Although SampleRate starts to adopt higher rates and increase the throughput from 20 s, it is unable to fully utilize the STP phase as the train is about to depart.

During the DEP phase, the performances of AARA and SampleRate are similar. In fact, SampleRate is a sensitive rate adaptation scheme, because it is conservative to raise its bit-rate, yet it is susceptible to drop off its bit-rate. In the DEP phase, the channel quality degrades quickly. SampleRate with default parameter is able to handle the degradation of channel quality because it can quickly decrease the bit-rate. Thus, we adjust the probing rate back to default value 10%. Besides, AARA shuts down the wireless device to enter sleep mode for the sake of conserving power during the CRU phase. We analyze the performance of power-saving in Section "Power saving".

To further investigate what causes the difference between the performances of AARA and SampleRate, we also analyze the ratio of transmitted packet in each bit-rate of all four phases. In Use Figure11, darker marked colors represent higher bit-rates. We see that AARA adopts more "higher rates" than SampleRate does during the most critical period, the STP phase. This is the main reason why the throughput of AARA outperforms that of SampleRate. Note that, in ARR phase, AARA uses much more higher bit-rates in HC-MRT compared to MC-MRT system. As mentioned earlier, the duration of ARR phase is shorter in MC-MRT, so the time AARA can use higher bit-rate is also shorter. Hence, the advantage of employing the bit-rate from the lowest rate in ARR phase is less noticeable in MC-MRT. Nevertheless, AARA still sends more packets in higher bit-rate in MC-MRT. In DEP phase, since the channel qualities degrade rapidly, both AARA and SampleRate tend to employ lower bit-rates. SampleRate uses more higher bit-rate in CRR phase, but the transmitted packets are mostly useless. This is because the number of packets in this phase is extremely few and the packets cannot be received due to the long distance. We also compute the ratio of transmitted packets in all the phases. The result shows that AARA transmits more packets in high bit-rate when the channel quality is good, and thereby AARA achieves higher throughput than SampleRate does.

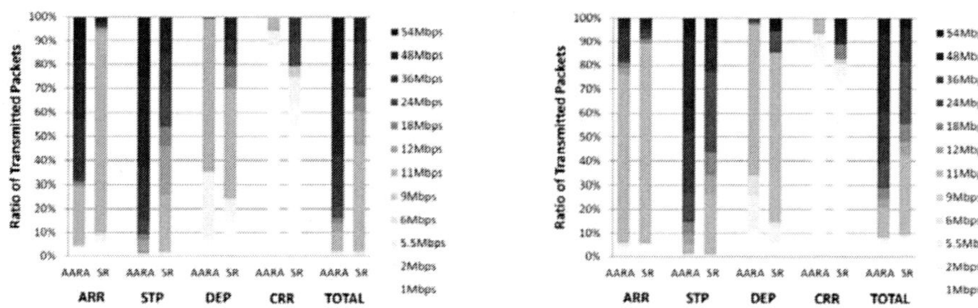

(a) High-Capacity MRT system (b) Medium-Capacity MRT system

Figure 11. THE RATIO OF TRANSMITTED PACKET IN EACH PHASES.

Next, we discuss the latency issue in AARA's phase estimation. The estimation of the phases is sequential. For instance, when in the ARR phase, the only two next possibilities are either to switch to the STP phase or to stay in the ARR phase. Because the most important phase to the overall throughput is STP phase, we focus on the estimation of STP phase start point. If the estimated start point is delayed, AARA will clear the history record at a later point. For example, if AARA estimates the train stopping in the station at a time that is already 5 s later than it really is, then AARA clears the history record 5 s later. This means we will waste 5 s without taking the opportunity to use higher bit-rates. However, such loss will not cause disaster as it only degrades the throughput gain. The overall throughput of AARA is still superior to SampleRate, which has no extra context information.

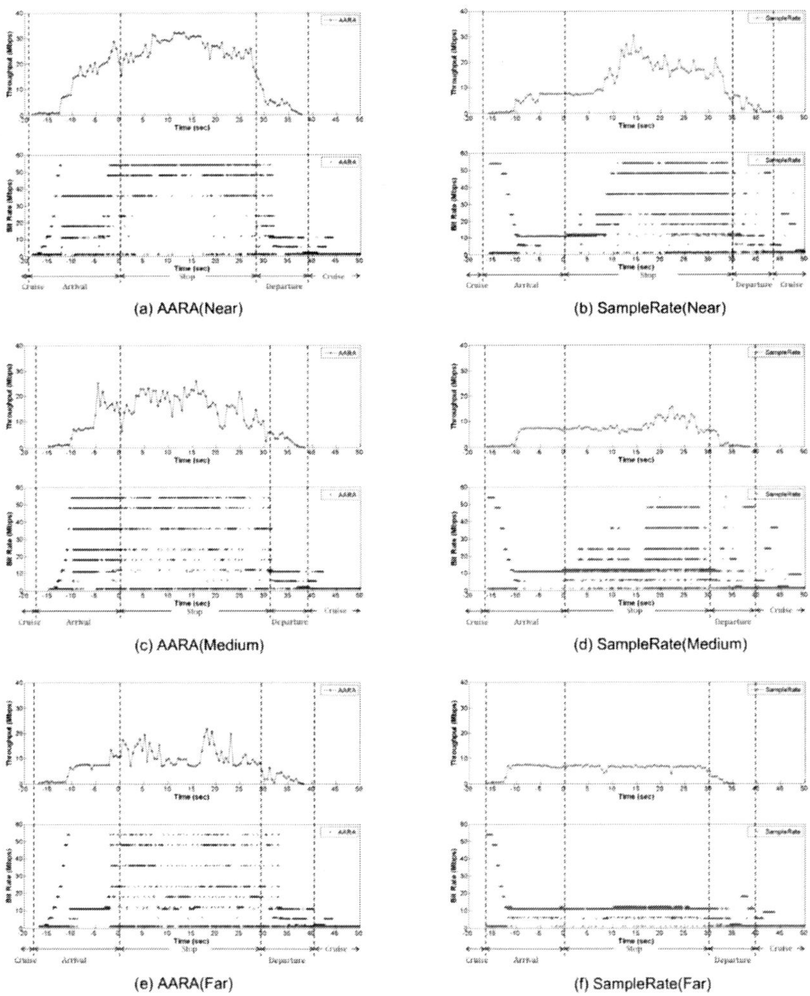

(a) AARA(Near)

(b) SampleRate(Near)

(c) AARA(Medium)

(d) SampleRate(Medium)

(e) AARA(Far)

(f) SampleRate(Far)

Figure 12. THE EFFECT OF CHANNEL QUALITY ON THE THROUGHPUT AND THE RATE-SELECTION OF AARA AND SAMPLERATE ALONG THE PATH IN HC-MRT. We conduct the experiments in three distances between the MS and BS. The RSSI values measured at the receiver when the train stops in the station are −46.30, −63.19, and −68.07 dBm as the scenarios of near, medium and far distance, respectively.

Effect of channel quality

So far, we focus on the idea that employing higher bit-rate to bring higher throughput when the channel quality is good. We also need to study that what if the channel quality is not sufficient to employ high bit-rate even when the channel quality is already the best. In this section, we evaluate the effect of channel quality as well. We design three scenarios of the experiments where the MS is placed at different positions in the car while keeping the BS in the same position on the platform. A total of three different MS-to-BS distances are measured (near, medium, far), as shown in Figure12. The distance of MS and BS correlates with the channel quality. The setting of the experiment of medium distance is the same as the previous subsection, so Figure 12c,d are the same as Figure 9a,b. We put the figures here again for easier comparison.

In the far scenario, because the transmission failure is too serious due to the worse channel quality, SampleRate does not try higher bit-rate at all; while AARA still tries higher bit-rate in STP phase. The main reason is that AARA cleans the stale transmission record at the beginning of STP phase, and also disable the blocking mechanism of bit-rate with consecutive failures. Though the channel quality is worse than that in the setting of medium distance, it is possible to send packets successfully in higher bit-rate sometimes. However, the sensitive property of SampleRate almost completely abandons the opportunities. On the other hand, in the near scenario, the overall throughput of AARA is naturally superior to that in the medium scenario. SampleRate in near scenario outperforms itself in medium scenario as well. The result is reasonable. It becomes easier for SampleRate to perceive the upgrade of the channel quality when the train enters STP phase since there is not so many accumulative transmission failure in the ARR phase. Thus, the higher bit-rate are not be blocked for so long. In other words, the channel qualities in the latter part of ARR phase and the STP phase are more similar, and thus the history record is partially credible than that in the medium distance scenario. Nevertheless, SampleRate still needs about 10 s after the train stops in the station to raise the bit-rate. In this scenario, AARA also boosts the bit-rate earlier than it does in the medium scenario. After testing the bit-rate from the lowest one in ARR phase, AARA senses it is able to use higher rate promptly, and thus boosts the overall throughput.

Figure13 depicts the overall throughput of AARA and SampleRate in each scenario. In HC-MRT system, the result show that the throughput decrease as the distance between the MS and the BS getting further. The throughput of AARA outperforms that of SampleRate in each scenario, even when the channel quality is worse (fewer chance to employ higher bit-rate) and better (SampleRate is able to perceive the channel upgrade more quickly). Reasonably, the improvement of the performance is most remarkable in the medium distance scenario. The throughput of AARA almost doubles that of SampleRate in this scenario.

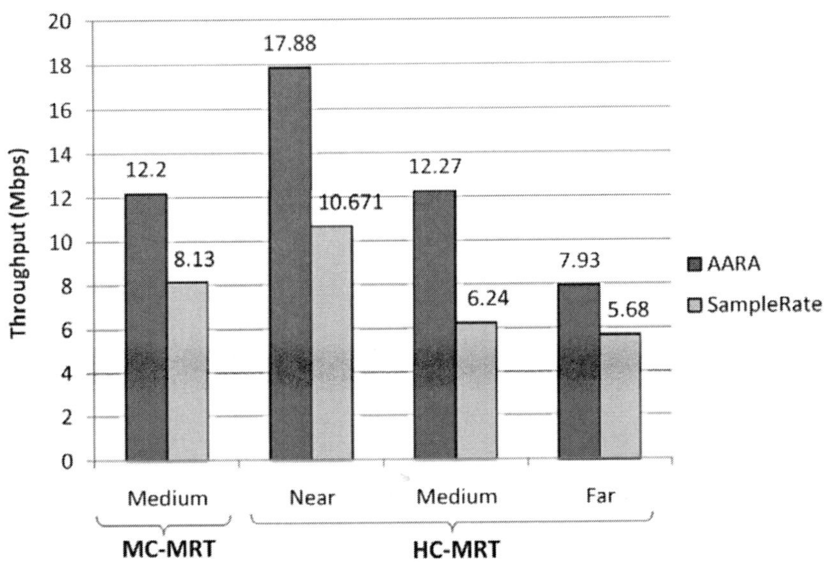

Figure 13. THE TOTAL THROUGHPUT IN DIFFERENT SCENARIOS.

The overall throughput gain in Figure13 is 1.50×in MC-MRT, 1.68× in the near distance scenario in HC-MRT, 1.97× in the medium distance scenario in HC-MRT, and 1.40×in the far distance scenario in HC-MRT. These experiments show that AARA, which is an enhancement of SampleRate according to out-of-band information, naturally performs better than that of SampleRate. Even when the channel quality is extremely bad (worse than our far scenario), AARA's throughput might only degrade to be almost as low as SampleRate's throughput.

Power saving

In addition to transmission performance, power saving is also deemed as a key issue. When the train departs from the platform, the MS and the BS are disconnected to each other. Leaving the wireless devices activated during this time consumes a substantial amount of energy. By contrast, if the wireless device is turned to sleep mode when data is not able to be transmitted, we can efficiently reduce power consumption. To achieve that, the MS has to automatically wake up the wireless device from sleep mode when the train is approaching to the next station. With the proposed mechanism of the train's movement estimation, it is easy to properly switch to sleep mode to save power, and to reactivate the device to transmit data. Specifically, we simply switch the wireless device into sleep mode at the beginning of CRU phase and reactivate it in the end of CRU phase. With the accurate phase estimation, AARA is able to reduce substantial power consumption without degrading the throughput.

We compute the amount of power AARA saves and show the experimental result in Figure14. The duration for calculating power consumption is from the beginning of DEP phase when a passenger gets on the train, to the end of ARR phase when the train stops at the station after the next station. For example, in Figure 4b, a passenger gets on the train at the MRT Wanfang Hospital station, moves toward MRT Wanfang Community station, and finally gets off at the MRT Muzha station. The BS is settled in Wanfang Community station. We compute the energy consumption from the beginning of DEP phase in MRT Wanfang Hospital station to the end of ARR phase in MRT Muzha station. There are three operating modes in commodity wireless cards: transmit mode, standby mode, and sleep mode. Transmit mode denotes the wireless card is activated and is transmitting the packets; standby mode denotes the wireless card is also activated but not transmitting packets; sleep mode denotes the wireless card is not activated. If power saving mechanism is not included, the wireless adaptor only operates in transmit mode and standby mode during a passenger's entire trip. According to the data sheet of NIC we used[23], transmit mode and standby mode will consume about 1,650 and 1,250 mW, respectively. Given the power saving mechanism, in CRU phase, we switch the wireless adaptor to sleep mode, which only consumes 70 mW. In fact, the power consumption of accelerometer is fairly low to be negligible.

Figure14 compares the energy consumption between AARA (with power saving) and default setting (without power saving) under the same condition stated above. In the experiments performed in HC-MRT and MC-MRT, AARA consumes, respectively only 54.2 and 46.6% of the total energy consumed by default settings. We see that the energy consumption in transmit mode is almost the same in AARA and default setting. However, AARA reduces the inessential power consumption in standby mode and increase the time operating the wireless device in sleep mode because the power consumed in sleep mode is much lower than in standby mode.

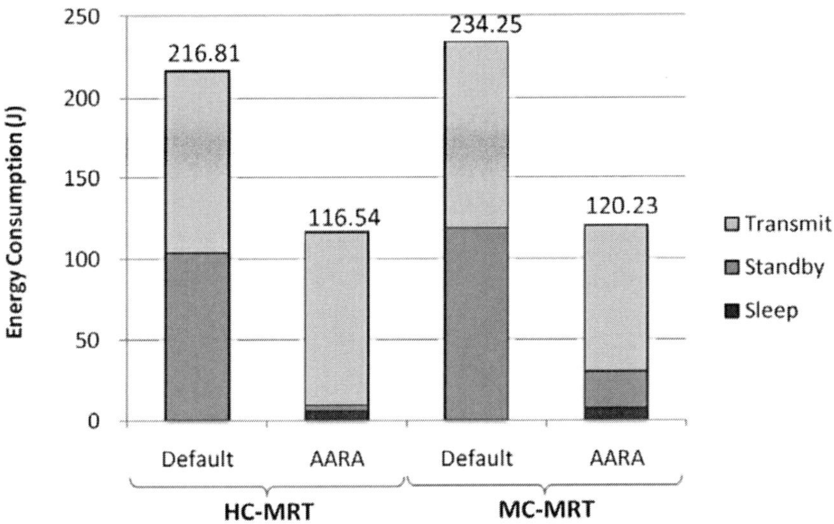

Figure 14. THE ENERGY CONSUMPTION OF DEFAULT SETTINGS AND AARA IN HC-MRT AND MC-MRT SYSTEMS.

Figure15 takes both energy consumption and transmission performance into consideration in the meanwhile. We compare the issue of energy consumption per bit in three experimental scenarios: default setting (SampleRate without power saving), AARA without and with the power saving mechanism. Here AARA without power saving mechanism only activates part of AARA's function—rate adaptation—but its function of shutting down the wireless card during CRU phase is disabled. This figure reveals that AARA's rate adaption component reduces the energy consumption per bit in terms of transmission performance. On the other hand, AARA's power saving component reduces the energy consumption per bit in terms of consuming less energy. The complete version of AARA consumes only 28.6 and 36.4% of the energy to transmit a data bit compared with the default setting in either MRT system.

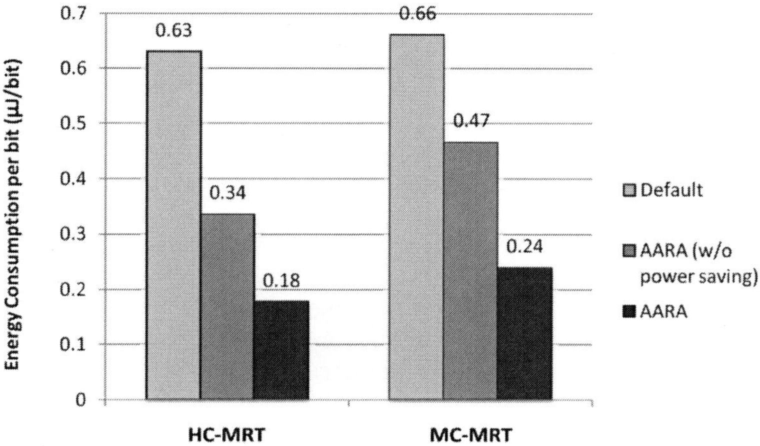

Figure 15. THE ENERGY CONSUMPTION PER BIT OF DEFAULT SETTINGS, AARA WITHOUT THE COMPONENT OF POWER SAVING, AND COMPLETE VERSION OF AARA IN HC-MRT AND MC-MRT SYSTEMS.

DISCUSSION

In this work, we have designed and implemented the AARA in mobile WiFi devices. The experimental results demonstrated the feasibility of the proposed scheme. In our current prototyping, the accelerometer is attached to the mobile device. As acceleration information has been shown useful to assist wireless communications, we could seek to obtain acceleration information with various methods. For example, for radio system built-in for train vehicles, the acceleration information might possibly obtained from train control system. For radio system on motorcycles, the acceleration information might be obtained from its control panel. Accelerometer could be built-in internally (e.g. smartphones have built-in accelerometer) or connected externally (e.g. connected via USB port). We foresee this acceleration-assisted wireless communication technology to be applied in different forms.

CONCLUSION

In this article, we develop a new rate adaptation mechanism, AARA, to enhance SampleRate to adapt the transmission rate with the aid of accelerometer. AARA achieves the goal of boosting transmission performance on the MRT system. We use accelerometer to identify the patterns of the vibration and movement of the train, and to estimate the four movement phases of the train, regardless of the accelerometer's orientations. Afterwards, we utilize the estimation result to employ suitable bit-rate adaptation strategies in each phases. Crucially, as STP phase provides the best transmission opportunity, AARA attempts to properly employ higher bit-rates so as to seize the limited time to maximize the throughput. In addition, we take power efficiency for transmission into concern. The experiments conducted on the two types of Taipei MRT system, High-Capacity MRT and Medium-Capacity MRT, show that AARA improves both the transmission performance and the power efficiency compared with the default settings.

ACKNOWLEDGEMENTS

This work was also supported by the National Science Council, the National Taiwan University and the Intel Corporation under GrantsNSC 100-2911-I-002-001, and 101R7501.

AUTHORS' AFFILIATIONS

(1) Department of Electrical Engineering, National Taiwan University

(2) Department of Electrical Engineering, Graduate Institute of Communication Engineering, Intel-NTU Connected Context Computing Center, National Taiwan University

REFERENCES

1. Zheng Y, Li Q, Chen Y, Xie X, Ma WY: Proceedings of the 10th international conference on Ubiquitous computing (UbiComp), 2008. Understanding mobility based on GPS data, pp. 312–321
2. Patterson D, Liao L, Fox D, Kautz H: Proceedings of the fifth international conference on Ubiquitous computing (UbiComp), 2003. Inferring high-levelbehavior from low-level sensors, pp. 73–89

3. Reddy S, Mun M, Burke J, Estrin D, Hansen M, Srivastava M: Using mobile phones to determine transportation modes,. ACM Transactions on Sensor Networks (TOSN) 2010, vol. 6(no. 2):1-27.
4. Lacage M, Manshaei M, Turletti T: Proceedings of the 7th ACM international symposium on Modeling, analysis and simulation of wireless and mobile systems (MSWiM), 2004. IEEE 802.11 rate adaptation: a practical approach, pp. 126–134
5. Onoe Rate Control [http://madwifi-project.org/wiki/UserDocs/RateControl]http://madwifi-project.org/wiki/UserDocs/RateControl []
6. Wong SHY, Yang H, Lu S, Bharghavan V: Proceedings of the 12th ACM international conference on Mobile computing and networking (MobiCom), 2006. Robust rate adaptation for 802.11 wireless networks, pp. 146–157
7. Bicket JC: Bit-rate Selection in Wireless Networks. Master's thesis, MIT, 2005
8. Shankar P, Nadeem T, Rosca J, Iftode L: IEEE International Conference on Network Protocols (ICNP), 2008. CARS: Context-aware rate selection for vehicular networks, pp. 1–12
9. Rahmati A, Zhong L: Proceedings of the fifth international conference on Mobile systems, applications, and services (MobiSys), 2007. Context-for-wireless: context-sensitive energy-efficient wireless datatransfer, pp. 165–178
10. Joshi A, Kulkarni P: Proceedings of the ACM SIGCOMM 2010 conference, Poster Session, 2010. Vehicular wifi access and rate adaptation, pp. 423–424
11. Santhapuri N, Manweiler J, Sen S, Bao X, Choudhury R, Nelakuditi S: 17th IEEE Workshop on Local and Metropolitan Area Networks (LANMAN), 2010. Sensor assisted wireless communication, pp. 1–5
12. Ott J, Kutscher D: Proceedings of the 23rd Conference of the IEEE Communications Society (INFOCOM), 2004, vol. 1. Drive-thru Internet: IEEE 802.11b for "automobile" users,
13. Hadaller D, Keshav S, Brecht T, Agarwal S: Proceedings of the fifth international conference on Mobile systems, applications and services (MobiSys), 2007. Vehicular opportunistic communication under the microscope, pp. 206 219
14. Eriksson J, Madden S: Proceedings of the 14th ACM international conference on Mobile computing and networking (MobiCom), 2008. Cabernet: vehicular content delivery using WiFi, pp. pp. 199–210

15. Zhao J, Arnold T, Zhang Y, Cao G: Proceedings of the fifth ACM international workshop on VehiculAr Inter-NETworking (VANET), 2008. Extending drive-thru data access by vehicle-to-vehicle relay, pp. 66–75
16. Giannoulis A, Fiore M, Knightly E: Proceedings of the 6th international conference on Mobile systems, applications, and services (MobiSys), 2008. Supporting vehicular mobility in urban multi-hop wireless networks, pp. 54–66
17. Deshpande P, Kashyap A, Sung C, Das S: Proceeding of the 7th international conference on Mobile systems, applications, and services (MobiSys), 2009. Predictive methods for improved vehicular WiFi access, pp. 263–276
18. Multiband Atheros Driver For WiFi [http://madwifi-project.org] []
19. Iperf [http://sourceforge.net/projects/iperf] []
20. Wireshark [http://www.wireshark.org/] []
21. Radiotap [http://www.radiotap.org/] []
22. Taipei Metro [http://english.trtc.com.tw] []
23. Data Sheet of Dlink DWL-AG660

11

CHAPTER

A COGNITIVE QOS MANAGEMENT FRAMEWORK FOR WLANS

Mostafa Pakparvar*, David Plets, Emmeric Tanghe, Dirk Deschrijver, Wei Liu, Krishnan Chemmangat,

Ingrid Moerman, Tom Dhaene, Luc Martens and Wout Joseph

ABSTRACT

Due to the precipitous growth of wireless networks and the paucity of spectrum, more interference is imposed to the wireless terminals which constraints their performance. In order to preserve such performance degradation, this paper proposes a framework which uses cognitive radio techniques for quality of service (QoS) management of wireless local area networks (LANs). The framework incorporates radio environment maps as input to a cognitive decision engine that steers the network to optimize its QoS parameters such as throughput. A novel experimentally verified heuristic physical model is developed to predict and optimize the throughput of wireless terminals. The framework was applied to realistic stationary and time-variant interference scenarios where an average throughput gain of 344% was achieved in the stationary interference scenario and 70% to 183% was gained in the time-variant interference scenario.

Keywords Dynamic spectrum access Radio environment maps Cognitive radios Interference management

1 INTRODUCTION

Large-scale growth of wireless networks and spectrum scarcity are introducing more interference than ever. Intensive interference degrades wireless links and may jeopardize continuous connectivity and hence quality of service (QoS) offered to the users.

Cognitive radios (CRs) are becoming a tempting solution to tackle this type of spectrum over-utilization by introducing opportunistic usage of frequency bands that are not heavily occupied by licensed users [1]. The equal regulatory status of wireless terminals on the industrial scientific medical (ISM) band leaves no consideration of a primary user for various wireless networks. Therefore, interoperability of ISM band networks is becoming a key issue that must be solved.

Wireless networks are designed to tackle homogeneous intra/inter network interference by means of various medium access techniques. The efficiency of interference-tackling techniques of the present wireless technologies has been profoundly assessed in the literature. For instance in [2], a large number of experiments prove that the throughput of IEEE 802.11 wireless local area network (WLAN) is highly dependent on the traffic characteristics of other present Wi-Fi networks.

The objective of this paper is to validate a framework that uses cognitive radio techniques for QoS management of WLANs. Cognitive radio techniques are majorly studied and practiced for operation of secondary users in licensed bands while this framework focuses on deploying such techniques in the unlicensed ISM band. The framework builds on radio environment maps (REMs) that contain PHY spectrum information obtained from a set of environment sensing devices. An experimentally verified heuristic physical model is developed to predict and optimize the throughput of wireless terminals based on the information contained in the REMs. The model is used as the kernel of the cognitive decision engine (CDE).

The novelties presented in this paper are the following: (1) extendable real-time indoor radio environment maps: maps that can be redesigned for any indoor environment, and (2) an experimentally verified heuristic cognitive decision engine which is based on a physical model that predicts throughput as a function of measurable wireless LAN parameters.

The model is derived from a set of exploratory measurements and is validated with various practical measurements in a pseudo shielded experimental testbed building, w-iLab.t [3].

The rest of this paper is organized as follows. In Section 2, after reviewing the related works, the cognitive framework is outlined. Section 3 describes a set of exploratory measurements we performed to derive a heuristic physical model which servers as the kernel of the cognitive decision engine. In Section 4, the development of the heuristic model is elaborated, and in Section 5, we explain how this model is integrated to the cognitive decision engine. Section 6 presents two proof of concept scenarios for static and time variant interference. Section 7 demonstrates an application outlook of the framework. Section 8 presents the conclusions.

2 THE COGNITIVE FRAMEWORK

2.1 Related work

In the context of cognitive frameworks that address efficient interoperability of homo/heterogeneous wireless networks, spectrum monitoring and decision making are the central topics addressed by the authors in the current literature. REMs [4] play a key role for environment monitoring in many cognitive radio solutions. REMs represent an integrated database providing information such as spectrum availability, regulations, and also the degree of channel utilization [4]. In terms of spectrum utilization, REMs have been proposed to measure power spectral density (PSD) in order to determine the degree of spectrum utilization in a certain geographical area. These models are produced from a set of measurements and the application of spatial interpolation techniques that provide an estimate at locations that lack dedicated measurements [5]. In addition to PSD maps, other authors present channel gain maps that capture information about the propagation medium [6]. In [7], the authors propose using medium utilization as a metric to be included in the REMs dedicated for wireless LANs. In [8], REMs have been used to intelligently guide spectrum access for deployment of a prototype of a long-term evolution (LTE) system that opportunistically exploits the spectral white spaces in the upper ultra high frequency (UHF) TV bands.

The *decision making* mechanisms in the present literature exploit a multitude of artificial intelligence (AI) algorithms to derive proper decisions for operation of the cognitive wireless network. artificial neural networks (ANN) have been used for radio parameter adaptation in CR [9,

10]. The ANN determines radio parameters for given channel states with three optimization goals, including meeting the bit error rate (BER), maximizing the throughput, and minimizing the transmit power. In [11], it is proposed that the use of ANN characterizes the real-time achievable communication performance in CR. Since the characterization is based on runtime measurements, it provides a certain learning capability that can be exploited by the cognitive engine. The simulation results demonstrate good modeling accuracy and flexibility in various applications and scenarios.

Metaheuristics [12] are used for combinatorial optimization in which an optimal solution is sought over a discrete search space. Evolutionary algorithms/genetic algorithms (GAs), simulated annealing (SA), tabu search (TS), and ant colony optimization (ACO) are examples of metaheuristic algorithms. Among the various metaheuristic algorithms, the GA has been widely adopted to solve multiobjective optimization problems and to dynamically configure the CR in response to the changing wireless environment [13, 14]. For instance in [14], a software testbed for CR with the spectrum-sensing capability is implemented and a GA-based cognitive engine to optimize radio parameters for dynamic spectrum access is validated.

Apart from all the aforementioned AI methods, there are also methods in the literature that are based upon ranking the channels in order of their capacity. The ranking is done by characterizing the channel activities and making estimations of the capacity accordingly. For instance, in [15], the authors derive a *physical model* for the throughput of WiMax networks based on monitoring physical layer parameter carrier to interference and noise ratio (CINR). In particular, the authors in [16] propose a spectrum decision framework for cognitive radio networks which addresses QoS management of the secondary users in response to certain events such as appearance of a primary user or degradation of the QoS. Thus, their proposed framework not only accounts for consideration of primary users but also maintains the QoS delivered to the secondary users by making spectrum decision according to the channel activities. The current paper is therefore a logical continuance and extension of [16] where we share a common architecture for tackling interference and optimizing the QoS. We extend the concept by bringing cognitive radio techniques for QoS management to the ISM band WLANs as well as incorporating radio environment maps to the framework. To this end, we have extended the framework by adding radio environment maps and a cognitive decision engine which utilizes a physical throughput model to estimate and optimize the throughput of the IEEE 802.11 network.

2.2 Setup architecture

The architecture of the cognitive framework is illustrated in Figure 1.

The environment comprises a number of nodes that belong to the network under test and a number of unknown terminals that are referred to as interference sources. The network under test will collect interference information and reports them along with its own state (all relevant parameters such as throughput) to the database (DB) (see Figure 1). This information needs to be monitored and processed in order to make proper decisions to optimize the QoS.

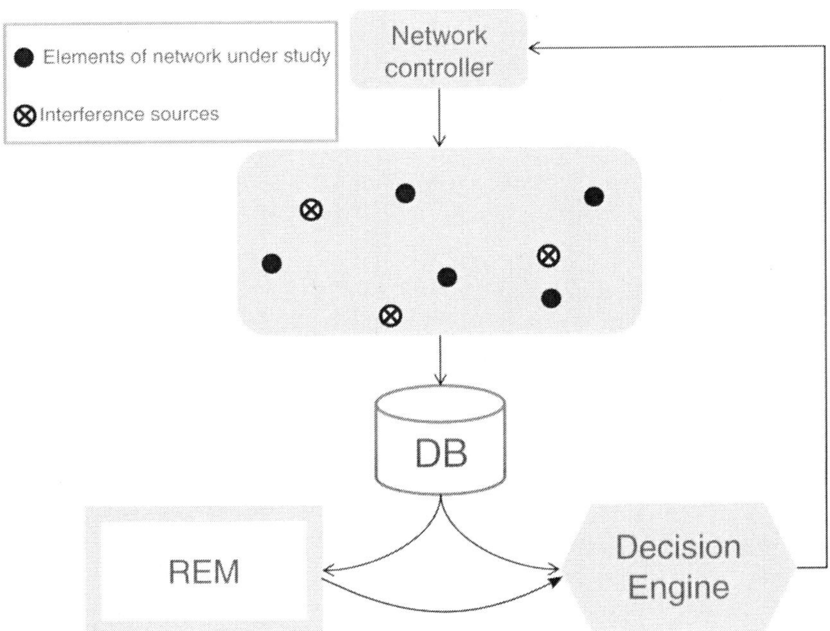

Figure 1. ARCHITECTURE OF THE FRAMEWORK.

2.3 Construction of the radio environment map

REMs [4] in this framework (Figure 1) refer to real-time spatial maps of the spectrum. These maps display vital information on spectral utilization across the whole network. As such, they facilitate environment monitoring by presenting relevant parameters that will be elaborated in Section 2.5. REMs can also be used by authorities to monitor spectrum for either finding violations of regulations or finding the so-called white spaces of the spectrum. For instance, apart from the real-life implementation of REMs in

[6], the IEEE 802.22 standard [17] recommends the authorities to prepare database services (similar to REMs) to provide public access to query for available TV channels. Indoor REMs give unprecedented insight on spectrum utilization in buildings with highly dense exploitation of wireless terminals.

Monitoring of spectrum in the wireless environment is a crucial part of any cognitive radio solution.

There are various solutions that enable wireless environment monitoring. Software-defined radios (SDRs) are advanced devices that are highly utilized within the cognitive radio networks for monitoring the intended bands [18]. Packet sniffers are an alternative solution used for monitoring the activity of a certain wireless technology on a certain band. Combining several packet sniffers on the different intended channels will bring the same functionality of a single software-defined radio device. The low cost and simplicity of implementation in case of the Wi-Fi technology is our motivation to only consider Wi-Fi packet sniffers as sources providing information of the spectrum.

The availability of information sources is restricted to a sparse set of (discrete) locations in the environment which necessitates utilizing interpolation techniques to estimate data values at arbitrary locations of the map. To this end, the architecture of the REM is designed as illustrated in Figure 2. The interpolation engine will extract raw data from the database in order to build an interpolation model. To preserve generality, the design foresees the feature for the user to select the appropriate interpolation technique according to their hardware and other requirements. The other input argument of the interpolation engine is the resolution of the interpolation that defines the number of points that must be interpolated based on the limited measured points. The interpolation method that is used in this paper by the REM is inverse distance weighting (IDW) [19]. This choice is motivated by the simplicity of underlying principle and the speed in calculation. The REM is implemented in the WiCa Heuristic Indoor Propagation Prediction tool [20].

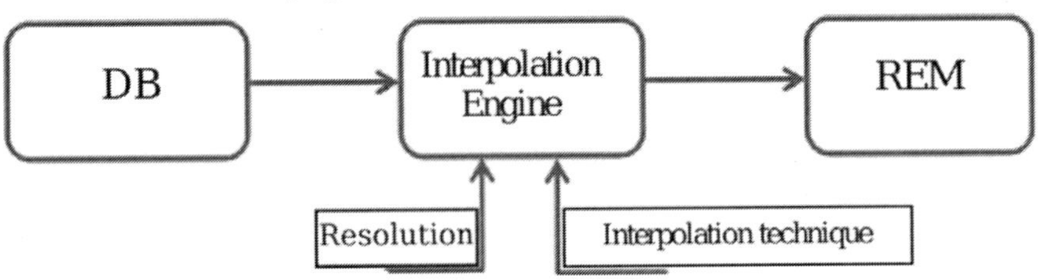

Figure 2. INTERPOLATION ARCHITECTURE.

2.4 The cognitive decision engine

The CDE in this framework (see Figure 1) refers to the entity that reconfigures the network with informed decisions that are based on its observation and cognition of the REMs. The architecture of the CDE is illustrated in Figure 3. The decision maker, in general, could be replaced by a human network manager or an automatic cognitive algorithm. The resource manager steers the network in response to the following events:

- Connection request: when a new connection is requested from the user.
- QoS degradation: when the clients of network under test report degraded quality of service.
- Sensing event: when the network under test is not active, this event keeps the decision maker informed of the channel quality on different bands.

The decision maker in this paper depends on a physical model that estimates the throughput based on interference observation. The decision maker invokes the models on the intended bands of operation and ranks the channels according to the estimations. The development of the physical model is further explained in Section 4.

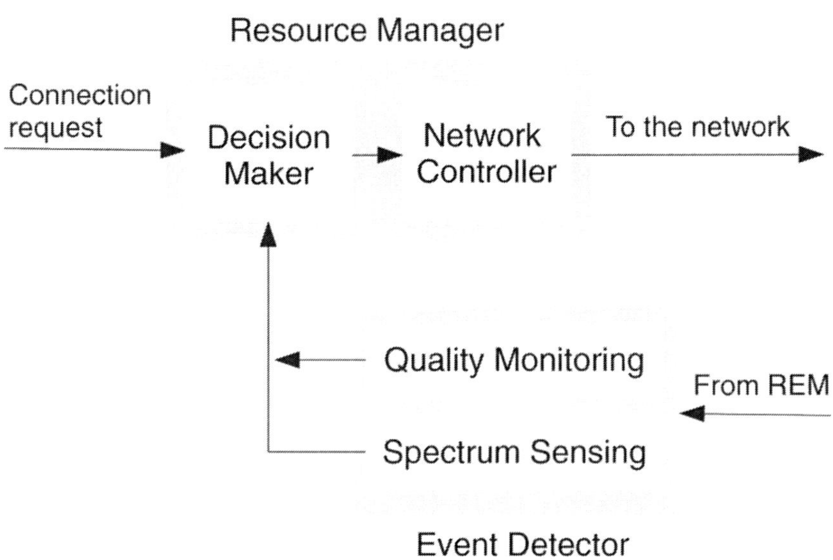

Figure 3. THE ARCHITECTURE OF THE COGNITIVE DECISION ENGINE.

2.5 Parameters

The different parameters that affect the performance of the wireless system can be classified as follows:

- Uncontrollable parameters: time-varying parameters that can be measured or monitored in the network but cannot be modified directly by the decision engine. They are further subdivided into two classes:
    - Target parameters (QoS/QoE): these are the parameters that the decision engine aims to optimize by taking corrective actions (e.g., throughput, latency, jitter, packet loss, mean opinion score (MOS)).
    - Measurable parameters (meters): these are the parameters that define the wireless environment and 'state' of the wireless system. These parameters are mostly informative (e.g., received signal strength indicator (RSSI) values, channel occupancy).
- Controllable parameters (knobs): configurable parameters that can be tuned or controlled by the cognitive decision engine (e.g., transmit power, packet size, location of the wireless entity).
-

The optimization process uses the knobs to optimize its target parameters given the input parameters (meters). This classification is shown in Table 1. Interference is the major issue that causes QoS degradation of wireless networks. Therefore, it should be characterized spatially, spectrally, and temporally in order to understand its influence on the QoS parameters. Table 1 also represents parameters that characterize interference. Parameters followed by a superscripted letter 'a' are those that are used in this research for deriving a model that predicts the throughput of a wireless network given the present interference characteristics.

Table 1. OPTIMIZATION PARAMETERS

Knobs (controllable)	Values	Unit	Meters (uncontrollable)	Unit	Target
Transmission			Network under test		QoS
Data generation rate	[0-54]	Mbps	Latency	ms	Latency
Transmission rate	[1,2,11,18,24,36,48,54]	Mbps	Jitter	ms	Jitter
Packet size	1,470	Bytes	Throughput (T)	Mbps	Throughput
Frequency channel	[1-13][a]		Packet loss	%	Packet loss
Power	[0-20][a]	dBm	Interference network		QoE
			Channel occupancy degree[a]	%	MOS
Location			Interference TxRate[a]	Mbps	
(x, y) coordinate		Meters	Interference power level[a]	dBm	
Height		Meters	Interference channel[a]		

[a]The parameters considered in the application of this paper.

Selection of the knobs in Table 1 is limited to the controllable parameters of IEEE 802.11g interfaces in the test environment which will be described in Section 3.1.1. Packet size is not considered since it is set primarily by the application and consequently controlled by the IP layer in a process called IP fragmentation to make sure it does not exceed the maximum transmission unit (MTU) [21]. The 1470 UDP packet size is used to avoid fragmentation of the packets.

Considering the audiovisual application of the network, instead of a QoS parameter, one may target a quality of experience (QoE) metric like mean opinion score (MOS) for audio or video streams. The audio MOS depends on QoS parameters, namely latency, jitter, and packet loss [22]. Finally, all meters are useful parameters to be represented on radio environment maps.

3 EXPLORATORY MEASUREMENTS

After describing the network configuration and the experiment setup, we will explain the exploratory measurement campaigns we performed for deriving the physical model of the throughput. Each measurement campaign aims at studying the influence of measurable interference characteristics on the throughput of the network under study.

3.1 Experiment description

3.1.1 Network configuration

All experiments are conducted in a pseudo-shielded testbed environment [3] in Ghent, Belgium. The nodes in the testbed are mounted in an open room (66 m × 20.5 m) in a grid configuration with an x-separation of 6 m and a y-separation of 3.6 m. Figure 4 shows the ground plan of the test lab with an indication of the location of the nodes. The 60 installed nodes are represented by the blue locations on the picture. Each node has two Wi-Fi interfaces (Sparklan WPEA-110N/E/11n mini PCIe 2T2R chipset: AR9280). Furthermore, a ZigBee sensor node and a USB 2.0 Bluetooth interface (Micro CI2v3.0 EDR) are incorporated into each node.

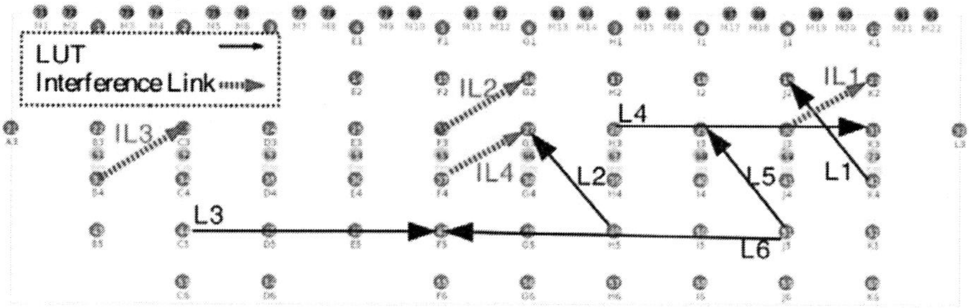

Figure 4. W-ILAB.T ENVIRONMENT. Links L1,L2,IL1,IL2,IL3,IL4 are 9.37 meters; L3,L4 are 18 meters; and L6 is 24 meters.

3.1.2 Experiment setup

For each of the interference measurement campaigns, a set of experiments is executed. Each experiment consists of two phases: a first phase where the QoS of the link under test is recorded without interference, and a second phase where interference is generated in the environment. Phase 2 of each experiment proceeds as follows: a. Start the interference transmission, b. wait 1 s to ensure interference is on the air, c. start the link-under-test transmission for *n* seconds, d. wait *n* seconds, and e. stop all data transmissions. The value of *n* specified for each measurement is mentioned at the corresponding measurement description. Data generation for the interference link and the link under test has been facilitated by the iperf traffic generator tool [23]. Using a client-server configuration, iperf outputs periodic reports of the throughput. Reports are parsed and stored in a database on the experiment controller server of the testbed.

During each interference measurement, two connection links are established, as displayed in Figure 4. The first link, always operating on channel 6, is denoted as the link under test (LUT). By default, this link is set at location L1 (see Figure 4) unless otherwise mentioned. The second link is the interfering link, between node 30 and node 20 (see IL1 in Figure 4). The data generation rate of the transmitters is defined as the User Datagram Protocol (UDP) bandwidth of the iperf application. All WLAN radio interfaces are set to transmit at a certain bit rate (TxRate), and iperf [23] was used as the traffic generator. The channel occupancy degree (COD) is here defined as the ratio of the data generation rate and the TxRate. There is a transmit buffer that is filled by data at a rate equal to the data generation rate, while the interface transmits through this buffer at a rate equal to the TxRate. The data generation rate is always smaller than or equal to the TxRate.

$$\text{COD}(\%) = \frac{\text{Data generation rate (Mbps)}}{\text{TxRate (Mbps)}} \times 100 \quad (1)$$

3.2 Measurement I: influence of the interference channel overlap, power level, and TxRate on the throughput

The first measurement studies the influence of channel overlap of the interference link on the throughput of the link under test. This measurement consists of 70 interference cases for every transmission rate of the interference link, each characterized by the interference channel (five channels) and by the generated interference transmission power (14 levels) as observed at node 19 (see link L1 in Figure 4) caused by interfering node 30 (see IL1 in Figure 4). The channel of the interfering link was varied from channel 6 (full overlap with link under test) to channel 10 (smallest overlap with link under test). For each channel overlap, 14 different interference power levels (dBm) were observed at node 19: [-51,-52,-53,-54,-56,-57,-59,-60,-62,-63,-66,-70,-71,-72].Translation of every specific transmission power of interfering node 30 into interference power levels at node 19 is done by averaging observations of the RSSI levels of packets transmitted from node 30 on the second Wi-Fi interface of the receiving node 19 which is operating on the monitor mode. This measurement was repeated for 2, 18, 36, and 54 Mbps interference transmission rates ($TxRate_{intf}$)to investigate if the $TxRate_{intf}$ influences the throughput on non-overlapping channels. In all cases, the transmission rate of the LUT is set to the maximum 802.11g rate of 54 Mbps and the COD of the interference link is equal to 100% (see Equation 1). The value of *n* (duration of the assessment) was set to 150 s in this measurement. Note that interference TxRate of 2 Mbps was not assessed on interference channels of 6 and 7 since the heavy spectrum utilization at 2 Mbps objects data transmission to an extent that iperf does not report any throughput value.

Figure 5 shows the achieved average throughput as a function of the received interference power level, for interference on different channels. A higher channel number corresponds with less overlap, i.e., for increasing channel separations, interference decreases. The interference level on the *x*-axis of this figure is defined as the power received from node 30 at node 19 (i.e., sender of IL1 to receiver of L1, see Figure 4).

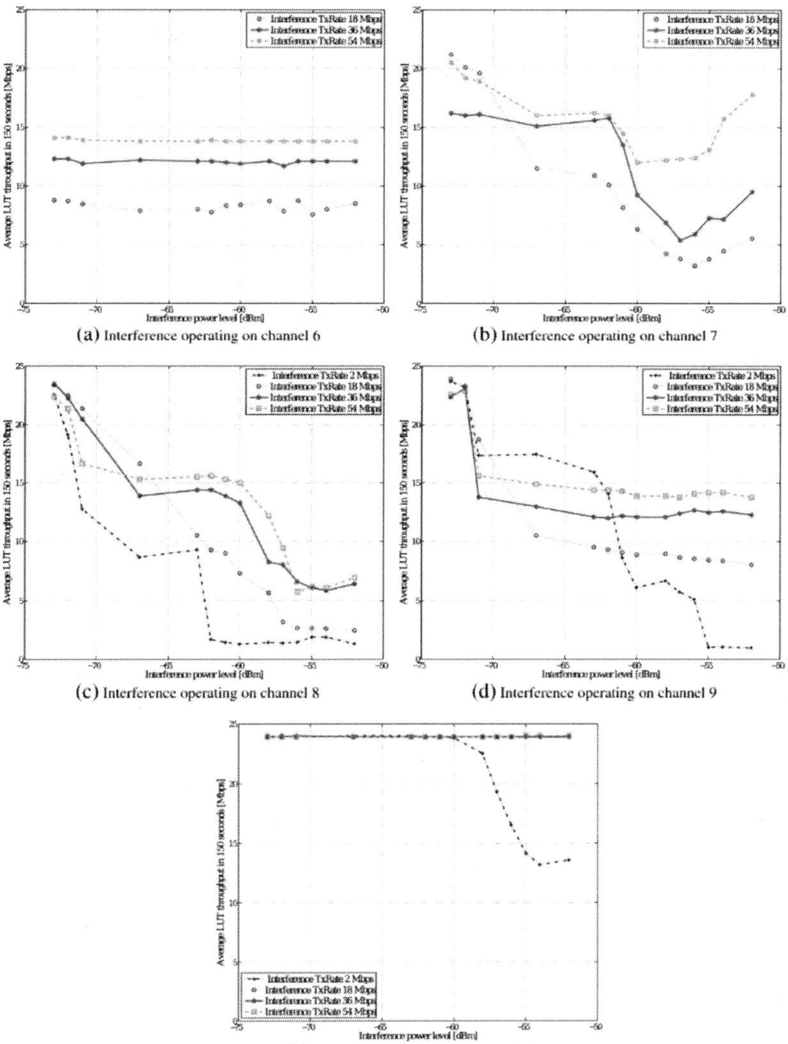

Figure 5. AVERAGE LUT THROUGHPUT FOR INTERFERENCE WITH DIFFERENT CHANNEL OVERLAPS. The LUT is on channel 6, interference on channel **(A)** 6. **(B)** 7. **(C)** 8. **(D)** 9 **(E)** 10.

Measurement results in Figure 5 show that the CMSA-CA mechanism of the IEEE 802.11g standard effectively tackles homogeneous interference present on the same channel. Despite the slight variations in the throughput values in Figure 5a, the order of the achieved throughput is preserved for increasing the interference power sensed at the receiver. The variations are the result of non-deterministic spectral competition between the interfering link and the link under test which is introduced by the random back-offs in the DCF algorithm of IEEE 802.11 standards [24]. Figure 5a also shows the dependency of the throughput to the interference TxRate. At lower TxRates, transmission of a fixed length packet lasts longer, causing the spectrum to be shared inefficiently between the interference link and the LUT (also reported in [25]). This dependency will be more elaborately studied and modeled in Section 3.4.

When the interference channel starts separating from the channel of LUT (Figure 5b channel 7 to Figure 5e channel 10), the channel sensing would become less accurate since only the common portion of the interference channel and LUT channel is sensed by the transmitter of LUT. Hence, the LUT transmitter is likely to sense a busy channel as free which introduces collision and throughput loss. This phenomenon becomes dominant as the channel separation is increased from 1 to 3 (interference on channels 7 to 9, Figure 5b,c,d, respectively). In this case, interference *power level* plays a key role due to the fact that higher interference signal power increases the probability of collision. Moreover, at worst case interference power level (-52 dBm), the throughput of LUT is higher when the *interference TxRate* is higher. This is also due to the more efficient spectrum sharing at higher interference TxRates.

When interference operates on channel 7 (Figure 5b), the LUT may recover if the interference power level is above -55 dBm, where the excess interference signal power compensates the missing portion of channel 7 that does not overlap with channel 6. This is apparent in Figure 5a where, for instance, the average throughput of the LUT in the presence of interference with TxRate of 54 Mbps changes from 21 to 13 and 18 Mbps at interference power levels of -72,-60, and -52 dBm, respectively. The other observation is when the interference is on channel 8 or 9 (Figure 5c,d), increasing the interference power level will not recover the throughput; in both cases, the throughput is dropped more than 20 Mbps when the interference power level is 20 dB increased for the interference TxRate of 2 Mbps.In Figure 5e, the throughput is dropped by a factor of 10 Mbps only for the 2-Mbps interference TxRate and the interference power levels above -60 dBm. This behavior is due to the longer spectrum occupancy of low TxRate packets that increases the probability of collision specially at higher interference signal powers. Figure 5e reveals that four-channel separation would avoid throughput loss (caused by collision) only if the interference transmission rate is high, i.e., the spectrum is utilized shorter for packet transmission.

Therefore, channel overlap between the interference source and the network under study is an important parameter that affects the throughput of WLANs. The results of these measurements suggest that in such conditions, if the cognitive decision engine has to select an overlapping channel, it should select the channel which is at least four channels apart (also recommended by the IEEE 802.11 standard [24]), and if two channels happen to be chosen, the one with the lowest interference power level and the highest interference TxRate should be selected.

3.3 Measurement II: influence of the interference COD on the throughput

The second measurement studies the influence of channel occupancy degree of the interference link on the throughput of the link under test. This measurement consists of 140 interference cases each characterized by a varying COD (10 levels) and a varying generated interference level (14 levels) at node 19 (caused by interfering node 30). The COD was varied from 10% to 100% in 10% steps. For each channel occupancy degree, 14 different interference power levels (dBm) were observed at node 19: [-51,-52,-53,-54,-56,-57,-59,-60,-62,-63,-66,-70,-71,-72].The data generation rate and transmit rate of the LUT were both set to 54 Mbps (COD = 100%). The transmit rate of the interfering link was set to 2 Mbps and was operated on channel 6, the same channel as the LUT. The value of *n* was set to 500 s.

Figure 6 shows the achieved average throughput (*T*) in 500 s for interference with varying channel occupancy degrees and with different interference power levels. As apparent from the figure, power levels do not play a key role in affecting the throughput. Analyzing the measurement results of the channel occupancy experiments implies an increasing nonlinear monotonic relation between the channel occupancy degree for COD values below 80%. This relation can be expressed by the model below

$$T(\mathrm{COD}_{\mathrm{Intf.}}) = ae^{-b.\mathrm{COD}_{\mathrm{Intf.}}}u(80 - \mathrm{COD}_{\mathrm{Intf.}}) + a^{-b*80}u(\mathrm{COD}_{\mathrm{Intf.}} - 80), \qquad (2)$$

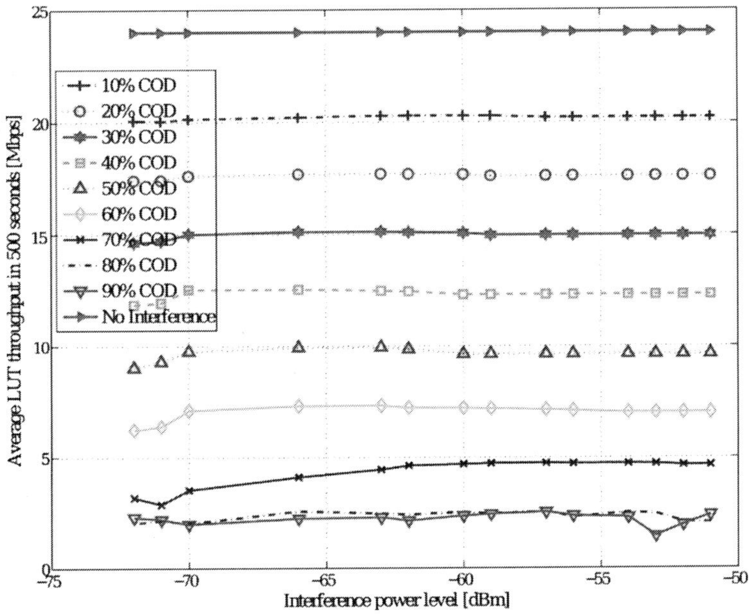

Figure 6. AVERAGE LUT THROUGHPUT UNDER THE ABSENCE OF INTERFERENCE AND FOR INTERFERENCE WITH DIFFERENT CODS.

where a=24.5, b=-0.023, and $COD_{Intf.}$ denote the interference COD. This model has a R^2 value of 0.97 and a root mean square error (RMSE) value of 1.18 Mbps. This fit has been calculated for throughput samples that are averaged over all interference power levels for every interference COD value. The statistics show an acceptable agreement between the model and the measurements which is also shown in Figure 7.

Channel occupancy degree indicates how the user is occupying the channel in time. Therefore, this metric can easily be calculated if the packet sniffer provides information on the transmission rate of the sniffed packets.

3.4 Measurement III: influence of the interference TxRate on the throughput

In the third interference measurement, the influence of interference transmit rate on the throughput over the link under test was assessed. This measurement consists of 98 interference cases each characterized by a varying interference transmit rate and a varying generated interference level (14 levels) at node 19 (caused by interfering node 30). The interference transmit rate was set to the values of [2,11,18,24,36,48,54] Mbps. For each interference transmit rate, 14 different interference power levels (dBm) were observed at node 19: [-51,-52,-53,-54,-56,-57,-59,-60,-62,-63,-66,-70,-71,-72].The data generation rate and transmit rate of the LUT were both set to 54 Mbps. The interfering link was operating on channel 6, the same channel as the LUT. The interference COD is always 100% in this measurement. The value of *n* in this measurement is set to 10 s.The measurement results in Figure 8 show a monotonic relation between the throughput and the interference transmission rate. Equation 3 models the behavior by an exponential polynomial:

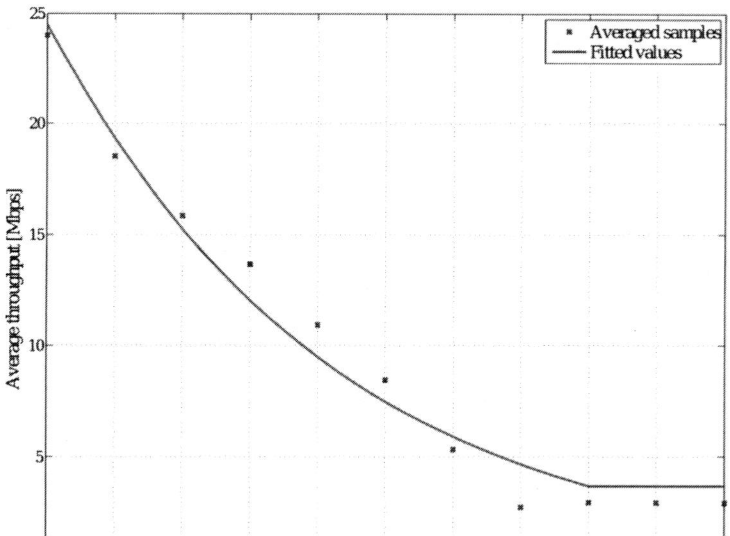

Figure 7. SAMPLE POINTS AND FITTED VALUES FOR MEASUREMENT II. Each point is an averaged value of the throughput measurement with a single interference COD value and the various interference power levels of Figure 6.

$$T(\text{TxRate}_{\text{Intf.}}) = ce^{d.\text{TxRate}_{\text{Intf.}}} + fe^{g.\text{TxRate}_{\text{Intf.}}}, \tag{3}$$

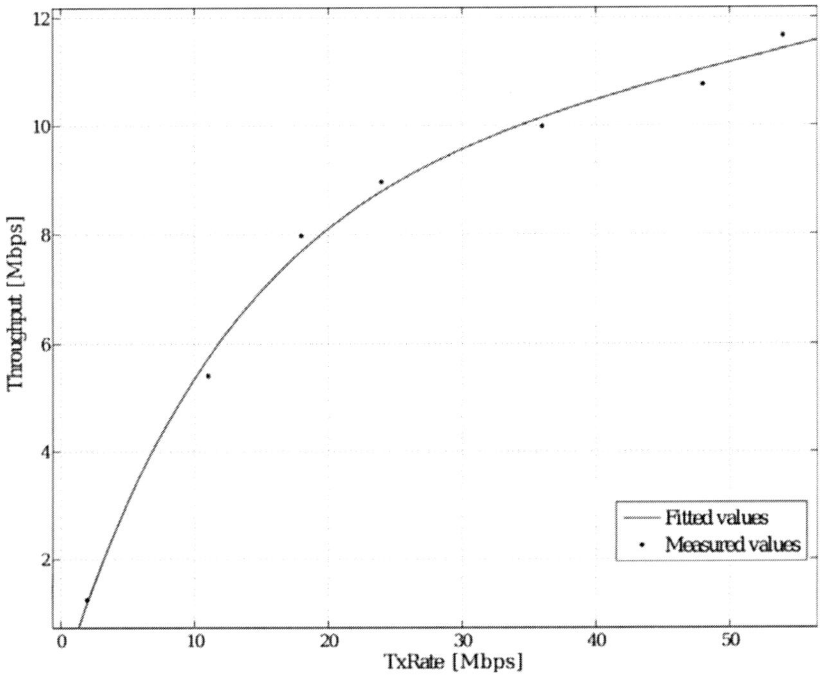

Figure 8. SAMPLE POINTS AND FITTED VALUES FOR MEASUREMENT III. Each point is an averaged value of throughput measurement with a single interference TxRate value and the various interference power levels described in the experiment.

where *c*=9.001, *d*=0.004, *f*=-9.344, *g*=-0.083, and $TxRate_{Intf.}$ denote interference TxRate (see Figure 8). This fit has an R^2 value of 0.9952 and an RMSE value of 0.3544 Mbps which is highly accurate.

3.5 Measurement IV: joint assessment of the interference TxRate and interference COD influence on the throughput

In measurements II and III, we assessed the influence of interference COD and TxRate on throughput by fixing one and varying the other parameter. This was a useful simplification to understand how throughput changes with any of those parameters. Since interference COD and interference TxRate are not completely uncorrelated parameters, deriving a decent two-dimensional model requires joint assessment of the two parameters influence on throughput. Here, link L3 serves as LUT and the same selection of interference link as described in previous measurements is used. We measure the average achievable throughput in a 10-s interval with interference TxRates of [2, 11, 18, 24, 36, 48, 54] and interference COD values starting from 0% to 100% in steps of 6.25% resulting in a total of 105 measurement cases. All links are set to channel 6, and the transmit power of all terminals is set to 20 dBm. The TxRate of the LUT is set to 54 Mbps, and its COD is set to 100% percent such that the maximum achievable throughput is measured. The value of *n* is again 10 s.Sample points in Figure 9 show the results of the measurement. Looking at the results of this measurement unveils the dependency of the interference COD and TxRate when predicting the throughput value. This is further investigated in the next section.

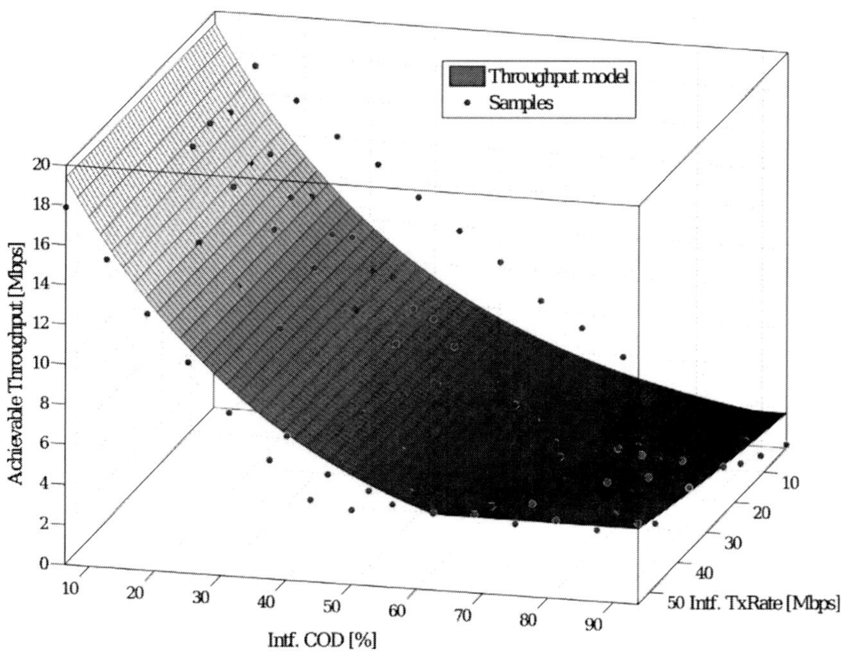

Figure 9. THE 3D DEMONSTRATION OF THE THROUGHPUT MODEL REALIZED ON LINK L3.

4 DEVELOPMENT OF THE MODEL

Data transmission rate of the Wi-Fi interface and the user activity are the key parameters that determine the spectrum utilization by any terminal. For a fixed amount of data transmission, higher transmission rate (TxRate) means less spectrum occupancy in time. Hence, an active user with low transmission rate occupies the spectrum to an extent that users of the other networks in the vicinity experience a much lower than expected throughput.

The model predicts the achievable throughput (*T*) based on the interference characteristics. The parameters that characterize the interference are the following: COD of the interference, interference TxRate, and interference frequency channel. All parameters are obtainable using the monitor mode of IEEE 802.11 interfaces [24] and packet tracing applications like libtrace [26] and tcpdump [27]. Extension to all Wi-Fi channels is feasible by periodic monitoring of all channels or by exploiting more WLAN interfaces at different locations of the network each operating on a single channel.

Due to the dependency of the interference parameters, deriving a decent two-dimensional model to predict the throughput is crucial.

From the results of measurements II and III, we know that there is an exponential behavior between the throughput (*T*) and interference parameters COD and TxRate when the interference is on the same channel as the LUT. The investigation of the throughput behavior in sample points of Figure 9 shows that depending on the interference TxRate, the achieved throughput increases exponentially below a certain interference COD threshold. Therefore, by applying unit step function to the exponential models and partitioning the parameters space into linear and exponential regions, we come up with an accurate model. This is evidenced by the results of our suggested model in Equation 4 below:

$$T(\mathrm{COD}_{\mathrm{Intf.}}, \mathrm{TxRate}_{\mathrm{Intf.}}) = (a_0 e^{-b.\mathrm{COD}_{\mathrm{Intf.}}}) u(90 - \mathrm{COD}_{\mathrm{Intf.}} - r.\mathrm{TxRate}_{\mathrm{Intf.}}) + \left(a_0 e^{-b(90 - r.\mathrm{TxRate}_{\mathrm{Intf.}})}\right) u(\mathrm{COD}_{\mathrm{Intf.}} + r.\mathrm{TxRate}_{\mathrm{Intf.}} - 90), \quad (4)$$

where $u(t)$ is the unit step function, $\mathrm{TxRate}_{\mathrm{Intf.}}$ denotes interference TxRate, $\mathrm{COD}_{\mathrm{Intf.}}$ denotes interference COD, and the coefficient parameters should be optimized for the intended link and the intended channel. The nonlinear least square optimization results for measurements on link L3 assign an a_0 value of 23.23, b=0.02 and r=0.5. These values are obtained by exploiting all the sample points of measurement IV. Figure 9 illustrates this realization of the model for sample points of link L3.The results of measurement IV show that the threshold of the aforementioned throughput exponential behavior decreases linearly from a COD of 90% for the TxRate of 2 Mbps up to the COD value of 60% for the TxRate of 54 Mbps. Inspired by this behavior, configuration of the step function arguments are optimized by a script that automatically investigates different combinations of the slope (*r*) and the intercept from a limited interval so that the least square error is achieved. The obtained R^2 value is 0.9425 and the RMSE value is 1.34 which show the accuracy of the model when used for the same link it was designed for. In the next section, we will show how the model is used inside the CDE and present its validation measurements.

The simplicity of this model enables the decision engine to devise it with a small number of sample points. Figure 10 shows the RMSE and R^2 of the model realized with varying number of sample points. Every realization is a set of measurements with all possible $TxRate_{Intf.}$ values and $COD_{Intf.}$ values uniformly distributed between 0% and 100%. The R^2 value is always more than 0.94 and the RMSE is always less than 1.35 Mbps. These statistics show that the model can be devised with 42 samples.

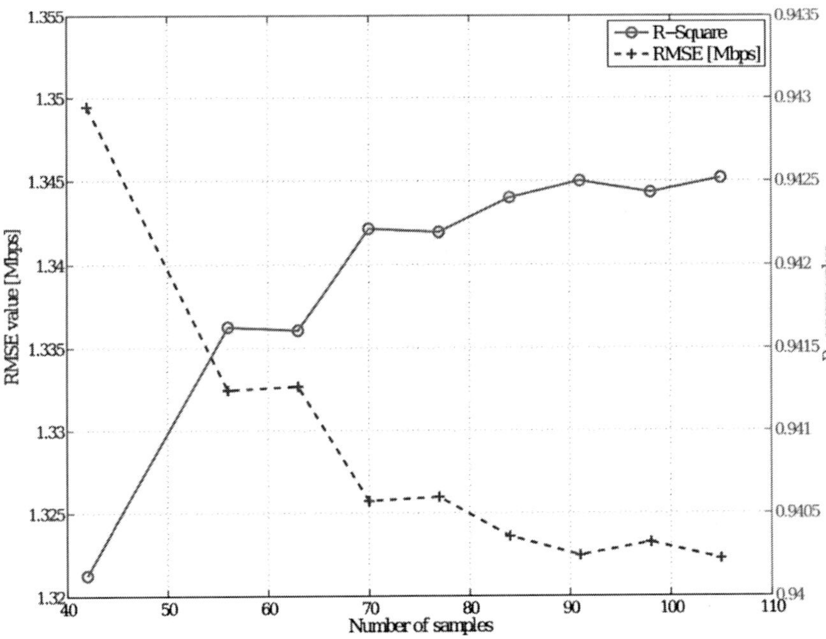

Figure 10. STATISTICS OF THE MODEL FOR REALIZATIONS WITH VARYING NUMBER OF SAMPLE POINTS.

5 VALIDATION AND INTEGRATION OF THE MODEL TO THE FRAMEWORK

In this section, we will show how the model developed in Section 4 will be used as the kernel of the cognitive decision engine. We will first explain how the input arguments of the model are prepared by the REM and will then describe how the CDE steers the network to the optimal configuration set. We will also present the validation measurements configuration and results in this section.

5.1 Monitoring of wireless environment

The integration of this physical model into a CDE requires measurement of its input arguments, $COD_{Intf.}$, and $TxRate_{Intf.}$. There are two possibilities to perform measurements over the whole Wi-Fi spectrum; either by utilizing a number of additional Wi-Fi terminals on the monitor mode each operating on a different Wi-Fi channel simultaneously or having the terminal sweep the channels periodically. In our implementation, we have used dedicated Wi-Fi interfaces as monitoring agents for every channel. The interfaces are brought up to the monitor mode of the IEEE 802.11 standard [24] to accept all ongoing detectable packets. A packet sniffer application [26] is then executed to capture all packets. The radio tap header [24] of every received packet provides crucial physical and link layer parameters such as the RSSI, packet length, and transmission rate of the packet. We propose to calculate the input arguments of the model of equation 4 by first calculating the equivalent interference transmission rate:

$$\text{TxRate}_{\text{eq.}}(\text{Mbps}) = \sum_{i \in N} R_i \frac{L_i}{L_T}, \tag{5}$$

where N is the total number of packets sniffed, R_i is the transmission rate of the sniffed packet, L_i is the length of the sniffed packet, and L_T is the total length of all sniffed packets in the current block of sniffed packets. Having the $TxRate_{eq.}$ calculated, we will find the equivalent interference COD by

$$\text{COD}_{\text{eq.}}(\%) = \frac{\frac{L_T(\text{Mb})}{\text{sniff interval (s)}}}{\text{TxRate}_{\text{eq.}}(\text{Mbps})} \times 100 \tag{6}$$

Using these relations, we can characterize the interference originating from several sources and give final values to the CDE for decision making. We should note that the COD estimations are accurate only if the sniffer captures the whole present Wi-Fi traffic without dropping any packets which is the case in our measurements. For coping with high data rates, we ensured zero packet drop by optimizing the script that was handling the tcpdump. This is realized by enlarging the tcpdump input buffer and dedicating more computational power to the corresponding process.

5.2 Decision making

Whenever a decision is requested from the CDE, it queries the REM for values of $COD_{Intf.}$ and $TxRate_{Intf.}$ for each channel. Once the inputs are ready, the CDE invokes the model on the channels with available inputs. The result is a vector of throughput estimations for every channel. Choosing the channel with the highest throughput value ignores the interference of overlapping channels. Therefore, if interference channels overlap with the monitored channels, the CDE should follow a rule-of-thumb approach where it takes into account both estimated throughput and the least interference influence from neighboring channels. The influence of interference from neighboring channels can be formulated by the results of measurement I, i.e., the CDE selects the channel with the least interference power level and the highest interference TxRate.

5.3 Model validation measurements

The objective of these measurements is to verify the two-dimensional model of Equation 4 for different sender-receiver distances, different link locations, different interference power levels, and different number of interference links. To this end, different combinations of LUT terminals with various distances of sender-receiver were used in combination with various number of interference links. The links used as LUT are illustrated in Figure 4: links L2 and L4 are replications of L1 and L3, respectively, at other locations. L5 and L6 are other links used as LUT in these measurements. There are also three interference links IL1, Il2, and IL3 evenly spaced among the testbed (see Figure 4). When there is more than one interference link present, the sniffer derives the $TxRate_{Intf.,eq.}$ and $COD_{Intf.,eq.}$ by using Equations 5 and 6,respectively.

For each location of the LUT, we investigate three cases for validating the model by utilizing one, two, and three interference sources simultaneously operating on the same channel of LUT. Each case is repeated with an interference transmission power of [0,10,14,20] dBm where we measure the average achievable throughput in a 10-s interval with interference TxRates of [2, 11, 18, 24, 36, 48, 54] and interference COD values starting from 0% to 100% in steps of 6.25%. All links are set to channel 6 and the transmit power of all terminals is set to 20 dBm. The TxRate of the LUT is set to 54 Mbps and its COD is set to 100% such that the maximum achievable throughput is measured.

5.4 Validation results

The model obtained by using measurements at link L3 is assessed with measurements on links L1, L2, L4, L5, and L6.

- The results of measurements with a single interference link are listed in Table 2.
- The results of measurements with more than one interference link are listed in Table 3.
-

Table 2. STATISTICS OF THE VERIFYING MEASUREMENTS WITH ONE INTERFERENCE LINK

LUT	Interference link	R^2	RMSE (Mbps)	Maximum deviation (Mbps)
L1	IL1	0.90	1.59	2.66
L2	IL1	0.90	1.59	2.57
L3	IL1	0.94	1.34	2.93
L4	IL1	0.89	1.65	3.31
L5	IL1	0.90	1.59	2.66
L5	IL2	0.90	1.59	2.57
L5	IL3	0.94	1.34	2.93

Table 3. STATISTICS OF THE VERIFYING MEASUREMENTS WITH MORE THAN ONE INTERFERENCE LINK

LUT	Interference link	R^2	RMSE (Mbps)	Maximum deviation (Mbps)
L5	IL1,IL2	0.89	1.46	2.61
L6	IL1,IL2	0.83	1.83	4.3
L5	IL1,IL2,IL3	0.83	1.62	+2.9
L6	IL1,IL2,IL3	0.84	1.83	2.7

The maximum deviation values of 2.9 Mbps for single interference link (Table 2) and 4.3 Mbps for multiple interference link (Table 3) might seem not ideal but having all R^2 values above 0.83 and the RMSEs not exceeding 1.9 Mbps. The CDE would safely use this model for comparing the performance of the link on different channels. One should be cautious that the objective of the CDE is to optimize the throughput rather than accurately estimating it. Hence, the model optimized for a single link location can serve to predict the throughput at other links. This measurement setup was limited to the line-of-sight node locations and open environment propagation characteristics of the testbed. The tolerance of the model to location variability of the interference within the line-of-sight environment is a good incentive to use it for similar environments like home and conference halls where the dominant propagation channel is line-of-sight. However, measurements to support this extension is the subject for future research. One the other hand, for environments with more diverse propagation characteristics and larger sender-receiver distances, the same methodology could be used to derive appropriate models for each type of environment.

6 PROOF OF CONCEPT

We investigate two scenarios in this proof of concept to show the relevance of integrating the physical throughput model into the CDE and also the efficiency of using the proposed framework to tackle interference and bring added value to the wireless network. In the first scenario, the CDE is used in a simplistic interference condition where interference characteristics are static, i.e., the $COD_{Intf.,eq.}$ and $TxRate_{Intf.,eq.}$ are stationary on every channel. In the second scenario, the interference characteristics on all channels are varied over time.

6.1 Static interference

As a static interference proof of concept, we will investigate the scenario consisting of three interference sources operating simultaneously on non-overlapping Wi-Fi channels 1, 6, and 11 each with a different COD and TxRate. Table 4 shows the interference profile of each source. For simplicity, in this proof of concept the CDE only seeks 2.4-GHz non-overlapping channels. The LUT is located on L3 (see Figure 4). The second Wi-Fi interface of the LUT transmitter scans the non-overlapping channels sequentially and quantifies the COD and TxRate of the interference on each channel. The CDE has to find the channel with the highest throughput estimation ($T_{est.}$) based on the proposed model. We also measured average throughput of LUT (T_{actual}) for each of the channels. The results are all listed in Table 4.

Table 4. CHARACTERISTICS OF THE INTERFERENCE SOURCES AND THE ESTIMATED AND ACTUAL ACHIEVED THROUGHPUTS ($T_{EST.}$ AND T_{ACTUAL} , RESPECTIVELY)

Channel	COD (%)	TxRate (Mbps)	$T_{est.}$ (Mbps)	T_{actual} (Mbps)
1	75	2	2.68	2.45
6	55	18	4.73	4.98
11	25	48	11.27	10.9

The LUT was initially operating on channel 1. The results show that if there was no CDE, the LUT had to survive with a throughput of 2.45 Mbps. This scenario shows that the CDE can steer the network to a better status by estimating the throughput on other channels. As such, by changing towards the channel proposed by the CDE (channel 11) the LUT can achieve a throughput of 10.9 Mbps which shows a 344% improvement. The results of this proof of concept also show the accuracy of the proposed model in estimating the throughput.

6.2 Time variant interference

In this section, we investigate a scenario with time variant interference profiles. The interference is varied over time by periodically changing the interference profile of each channel. At the beginning of each interval, the interference profile is varied by changing the number of interference links and their corresponding COD and TxRate on each channel and it is fixed until the end of the interval. We assume that only IEEE 802.11g channels 1 and 6 are available. The duration of the interval should be selected such that the sniffer would be capable of characterizing the interference on all the intended channels.

The duration of each interval in this scenario is set to 60 s firstly to ensure proper interference characterization and secondly to balance the interference dynamism. At the end of each interval, the CDE uses the collected information to steer the channel of LUT in order to optimize its throughput. The transition time depends majorly on the duration of the sniffing process. Lower sniffer intervals maintain smoother operation of the CDE. However, they are likely to lose interference profile changes if they are placed on the edge of profile change. We have used the CDE with two sniffer durations of 5 and 10 s. Table 5 shows the interference profile of each interval on each of the channels. In the first three intervals (0 to 180 s), there is a single interference link on any of the channels. During the intervals from 181 to 300 s, on one of the channels, there is interference from two links. In the last interval (301 to 360 s), channel 1 has interference from three interfering links and channel 6 has a single interfering link. For intervals in Table 5 where there is more than one interference link, the equivalent interference configuration is listed before individual link configurations in brackets. The LUT is located on L5 (see Figure 4). The second Wi-Fi interface of the LUT transmitter scans the channels 1 and 6 sequentially. The CDE has to find the channel with the highest throughput estimation ($T_{est.}$) based on the proposed model. We also measured average throughput of LUT (T_{actual}) in a separate measurement with the same interference profiles but without using the CDE to switch the channel of the LUT.

The results of the throughput measurements as well as the throughput estimations of the model are presented in Figure 11 for sniff interval of 5 s (Figure 11a) and 10 s (Figure 11b). This scenario shows that the CDE can steer the network to a better status by estimating the throughput on other channels. Depending on the sniffer interval, there might be transitional spikes in the CDE performance which could be solved by decreasing the sniffing period. As such, by changing towards the channel proposed by the CDE, the LUT can achieve an average throughput gain of 183% and 70%

compared to the case where it was constantly on channels 1 and 6, respectively. Figure 11 also shows that the throughput estimations follow the actual achieved throughput values on any of the channels despite the slight variations from the achieved throughput levels in intervals 120 to 180 and 240 to 300 which are nevertheless below the maximum deviation values obtained in the validation measurements in Section 5.4.

Table 5. CHARACTERISTICS OF THE INTERFERENCE SOURCES DURING EACH INTERVAL

	Channel 1			Channel 6		
Interval	Interference links	$COD_{eq.}$	$TxRate_{eq.}$	Interference links	$COD_{eq.}$	$TxRate_{eq.}$
0 to 60	IL1	10	11	IL2	35	36
61 to 120	IL3	60	48	IL1	10	18
121 to 180	IL2	35	11	IL3	10	54
181 to 240	[IL1, IL2]	20 [10, 10]	36 [36]	IL3	40	54
241 to 300	IL3	60	54	[IL1, IL2]	40 [20, 20]	18 [18, 18]
301 to 360	[IL1, IL2, IL3]	45 [15, 15, 15]	54 [54, 54, 54]	IL4	35	18

6.3 Discussion on the limitations

Firstly, the model is developed in a pseudo-shielded all line-of-sight single-hop environment where the measurements were performed. The hidden node problem is not taken into consideration since it is not an issue in this environment. More precisely, the model does not consider interference of nodes that are not in the radio range of the sniffers. The sniffers are assumed to capture all present traffic without dropping any packets. If for any reason (other ISM band interference such as Bluetooth, microwave ovens, and ZigBee) the interference is not perceived by the sniffers, the event detector of the CDE (see Figure 3) assists by detecting the unpredicted throughput degradation in the network.

Secondly, the physical model is developed for UDP traffic since its objective was to find the maximum capacity of the channel given a certain interference profile. If the application requires a reliable end-to-end connection, the same methodology we have incorporated in this paper shall be used to create a model which accounts for TCP traffic.

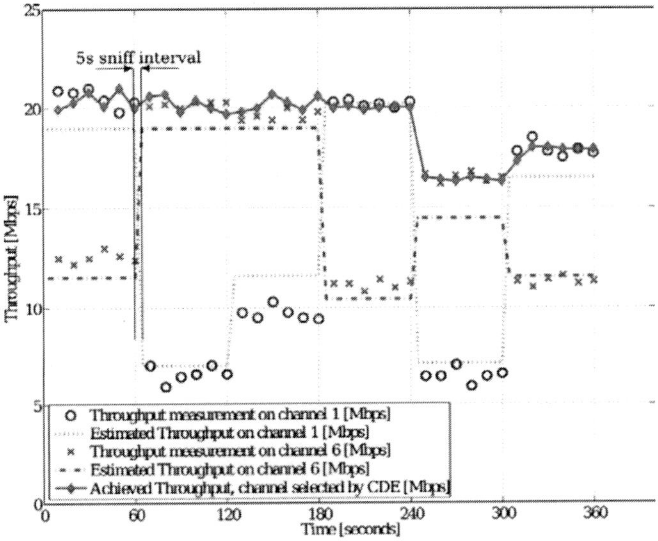

(a) 5 seconds sniff interval

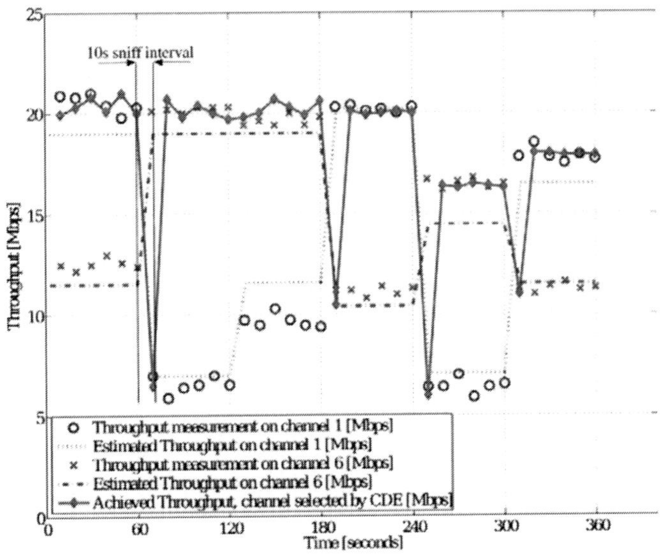

(b) 10 seconds sniff interval

Figure 11. COMPARISON OF THROUGHPUT ON CHANNELS 1 AND 6 WITH AND WITHOUT USING THE CDE. With sniff intervals of **(A)** 5 s and **(B)** 10 s.

Thirdly, deploying both the CDE and the REM imposes data and processing overhead on the network. Spectrum monitoring data overhead could be formulated in two cases depending on the mobility of spectrum sensing devices. For a distributed spectrum sensing which collects information from fixed and mobile terminals, the data overhead to update the REM would scale with the number of mobile terminals and update the frequency of the REM. Excluding mobile devices from spectrum monitoring and only depending on the fixed Wi-Fi sniffers do not introduce any data overhead to the wireless network. Event detection also imposes data overhead to the wireless network which would be negligible short-length data packets in the order of a few bytes for each event.

7 APPLICATION OUTLOOK OF THE FRAMEWORK

The goal of this section is to showcase the usefulness of the REMs and the proposed methodology in this cognitive framework when it is used for optimizing either throughput or audio quality in a multicast scenario. Here, the network under test (NUT) comprises a number of clients that are spread evenly among the w-iLab.t testbed. Depending on the target QoS metric (throughput or audio quality), the clients will listen to the audio stream or iperf multicast stream that is generated by the stream server that is at the same time the access point of the NUT. Interference networks are fixed in location although the interference profile, i.e., traffic characteristic of the interference, is adjustable.

When the user initiates the scenario, the audio/iperf stream server starts to transmit multicast audio/iperf traffic with an initial set of traffic configurations. If the target QoS metric is audio quality, at the same time, all clients will translate their received network characteristics (throughput, delay, and jitter) into audio MOS scores (in a scale of 1 to 5, where 1 corresponds to the poorest quality and 5 to the highest) and store the result in the DB. The MOS value is calculated by the E Model proposed by the ITU G.107 recommendation [22]. The REM will then illustrate the interpolated map of the MOS/throughput and other relevant parameters and feeds the necessary arguments to the CDE.

Due to the fundamental throughput-delay trade off [28], selecting the channel with the highest throughput estimation does not necessarily guarantee the lowest delay or jitter. Consequently, for audio quality target metric, the CDE should therefore use a different model that estimates the MOS value based on the estimated throughput and reported delay and jitter from the REM on each channel. In this way, the CDE estimates the MOS values on each channel and determines corrective actions that must be implemented on the NUT for the next round of audio streaming. Figure 12 is an example of REMs for RSSI and MOS values. Each map illustrates its corresponding parameter alongside the location of the network elements. Clients that are closer to the sender indicate higher RSSI values and consequently bring about higher MOS values to their users.

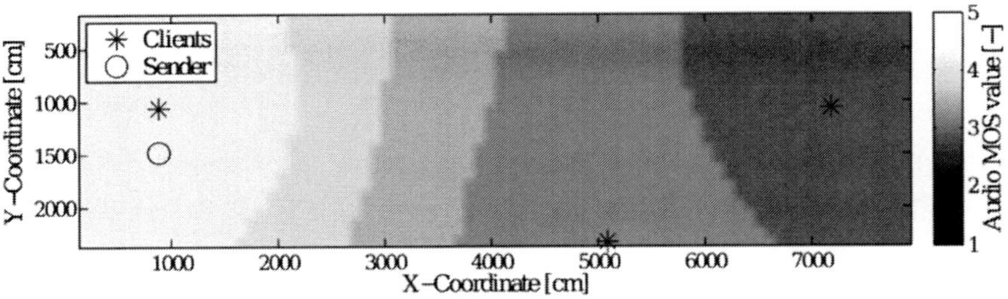

(a) REM of MOS values

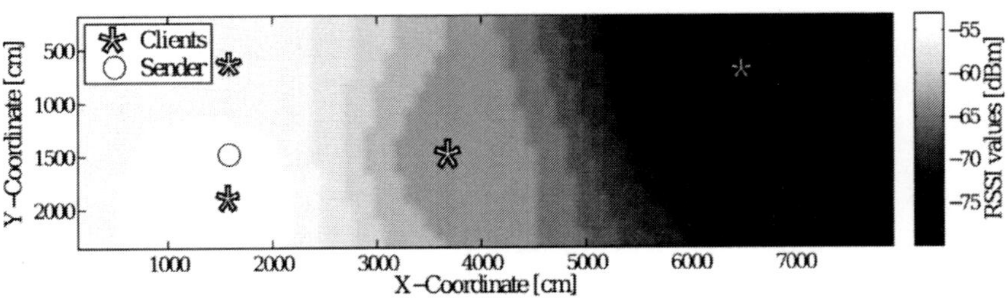

(b) REM of RSSI values

Figure 12. IDW INTERPOLATED REMS OF MOS AND RSSI VALUES ON THE W-ILAB.T TESTBED. (A) REM of MOS values. (B) REM of RSSI values.

When the corrective actions have been implemented and the MOS/throughput have been illustrated by the REM, we need to assess the efficiency of corrective actions, i.e., how much the network is performing better in terms of the intended metric (MOS/throughput). To address this, we define the objective of the scenario in the next paragraph.

The optimal configuration set of the network (node locations and all the parameters mentioned earlier) is achieved when $\forall i \in NUT(QoS_i \geq QoS_0)$. In this formula, QoS_i refers to the QoS value of the *i* th client of NUT and QoS_0 is a certain threshold. This scenario will follow the following steps in order to showcase the advantages of the REM and the CDEs.

1. All clients are set to the default (initial) configuration set.

2. While there is no interference in the environment, the NUT starts the first round of audio/iperf stream and all the clients record their MOS/throughput.
3. Based on the results of the previous round, the REM represents the QoS values and the CDE will optimize the configuration set by means of the corrective actions.
4. The interference network is set to its default profile. The interference stream starts here.
5. While there is interference on the air, the NUT starts another round of audio/iperf streaming. As before, all clients keep track of the QoS values.
6. At this step, the REM and/or the CDE should quantify the added value of the corrective actions in terms of ΔMOS, throughput, energy, etc.
7. If the objective is achieved, the iteration will stop; otherwise, CDE implements corrective actions to the NUT and the iteration starts again from 4.

This is a very simple example of how the models can be used. Future research will comprise the development of physical models for audio/video QoS metrics and a CDE based on the corresponding models.

8 CONCLUSION AND FUTURE WORK

A novel framework for cognitive wireless networking is proposed in this paper. The framework comprises different elements. REMs and the cognitive decision engine are the most integral parts of the framework to characterize the radio environment and make appropriate decisions reactively to the environment dynamism. Based upon numerous exploratory measurements, a novel physical throughput model accounting for interference channel occupancy degree and interference transmission rate was devised and verified in a pseudo-shielded testlab environment. The model was implemented within the CDE. The framework was applied to realistic stationary and time-variant interference scenarios where an average throughput gain of 344% was gained in the stationary interference scenario and 70% to 183% was gained in the time-variant interference scenario.

Future research on this framework includes comparison of the performance of the CDE based on the proposed model with smarter higher level algorithms and the integration of this type of physical modeling with other decision algorithms to achieve more efficient algorithms. The physical model was developed in a pseudo-shielded line-of-sight environment; therefore, on the one hand, extension to other line-of-sight environments like home or conference halls should be investigated, and on the other hand, incorporating the same methodology to develop similar models for other types of environments would be the topic of further research.

ACKNOWLEDGEMENTS

This work was supported by the iMinds-ICON QoCON project, co-funded by iMinds, a research institute founded by the Flemish Government in 2004, and the involved companies and institutions. DD is a Post-Doctoral Fellow of the FWO-V (Research Foundation - Flanders). The research is also partly funded by the Fund for Scientific Research - Flanders (FWO-V, Belgium) project G.0325.11N.

AUTHORS' CONTRIBUTIONS

MP executed all measurements and drafted the manuscript in collaboration with DP, WJ, and LM. DD, KC,and TD contributed partially to the radio environment maps by optimizing the IDW interpolation technique. WL and IM participated in the development of the passive packet sniffer. ET contributed in formulating the physical model and its statistical assessment. All authors read and approved the final manuscript.

AUTHORS' AFFILIATIONS

1. Department of Information Technology (INTEC), Ghent University-iMinds

REFERENCES

1. Mitola IJ, Maguire JGQ: Cognitive radio: making software radios more personal. IEEE Pers. Commun 1999, 6(4):13-18. doi:10.1109/98.788210 10.1109/98.788210
2. Plets D, Pakparvar M, Joseph W, Martens L: Influence of intra-network interference on quality of service in wireless LANs. In Proceedings of the IEEE International Symposium on Broadband Multimedia Systems and Broadcasting (BMSB). London, UK; 2013.
3. Bouckaert S, Becue P, Vermeulen B, Jooris B, Moerman I, Demeester P: Federating wired and wireless test facilities through Emulab and OMF: the iLab.t use case. In Proceedings of TridentCom. Thessaloniki, Greece; 2012.
4. Zhao Y: Enabling cognitive radios through radio environment maps. PhD thesis, Virginia Polytechnic Institute and State University, Blacksburg, Virginia. 2007.http://scholar.lib.vt.edu/ theses/available/etd-05212007-162735/ . Accessed 28 June 2013
5. Riihijarvi J, Mahonen P, Wellens M, Gordziel M: Characterization and modelling of spectrum for dynamic spectrum access with spatial statistics and random fields. In IEEE 19th International Symposium on Personal, Indoor and Mobile Radio Communications (PIMRC). Cannes, France; 2008:1-6. doi:10.1109/PIMRC.2008.4699912
6. Kim S-J, Dall'Anese E, Giannakis G: Cooperative spectrum sensing for cognitive radios using Kriged Kalman filtering. IEEE J. Selected Topics Signal Process 2011, 5: 24-36.
7. Portoles-Comeras M, Ibars C, Nunez-Martinez J, Mangues-Bafalluy J: Characterizing WLAN medium utilization for radio environment maps. In 2011 IEEE Vehicular Technology Conference (VTC Fall). San Francisco, CA, USA; 2011:1-5. doi:10.1109/VETECF.2011.6093284
8. van de Beek J, Lidstrom E, Cai T, Xie Y, Rakovic V, Atanasovski V, Gavrilovska L, Riihijarvi J, Mahonen P, Dejonghe A, Van Wesemael P, Desmet M: REM-enabled opportunistic, LTE in the TV band. In IEEE International Symposium on Dynamic Spectrum Access Networks (DYSPAN). Bellevue, WA, USA; 2012:272-273. doi:10.1109/DYSPAN.2012.6478140

9. Hasegawa M, Tran H-N, Miyamoto G, Murata Y, Kato S: Distributed optimization based on neurodynamics for cognitive wireless clouds. In IEEE 18th International Symposium on Personal, Indoor and Mobile Radio Communications (PIMRC). Athens, Greece; 2007:1-5. doi:10.1109/PIMRC.2007.4394658
10. Zhang Z, Xie X: Intelligent cognitive radio: research on learning and evaluation of CR based on neural network. In ITI 5th International Conference on Information and Communications Technology (ICICT). Cairo, Egypt; 2007:33-37. doi:10.1109/ITICT.2007.4475612
11. Baldo N, Zorzi M: Learning and adaptation in cognitive radios using neural networks. In 5th IEEE Consumer Communications and Networking Conference (CCNC). Las Vegas, NV, USA; 2008:998-1003. doi:10.1109/ccnc08.2007.229
12. Blum C, Roli A: Metaheuristics in combinatorial optimization: overview and conceptual comparison. ACM Comput. Surv 2003, 35(3):268-308. doi:10.1145/937503.937505 10.1145/937503.937505
13. Thilakawardana D, Moessner K: A genetic approach to cell-by-cell dynamic spectrum allocation for optimising spectral efficiency in wireless mobile systems. In 2nd International Conference on Cognitive Radio Oriented Wireless Networks and Communications (CrownCom). Orlando, FL, USA; 2007:367-372. doi:10.1109/CROWNCOM.2007.4549825
14. Kim JM, Sohn SH, Han N, Zheng G, Kim YM, Lee JK: Cognitive radio software testbed using dual optimization in genetic algorithm. In 3rd International Conference on Cognitive Radio Oriented Wireless Networks and Communications (CrownCom). Singapore; 2008:1-6. doi:10.1109/CROWNCOM.2008.4562484
15. De Bruyne J, Joseph W, Verloock L, Olivier C, De Ketelaere W, Martens L: Field measurements performance analysis of an 802.16 system in a suburban environment. Wireless Commun. IEEE Trans 2009, 8(3):1424-1434. doi:10.1109/TWC.2009.080085
16. Lee W-Y, Akyldiz IF: A spectrum decision framework for cognitive radio networks. Mobile Comput. IEEE Trans 2011, 10(2):161-174. doi:10.1109/TMC.2010.147
17. IEEE: IEEE Std 802.22-2012: IEEE Recommended Practice for Information Technology-Telecommunications and Information

Exchange Between Systems Wireless Regional Area Networks (WRAN)—Specific Requirements Part 22.2: Installation and Deployment of IEEE 802.22 TM Systems. IEEE, Piscataway; 2012.

18. Liu W, Pareit D, Poorter ED, Moerman I: Advanced spectrum sensing with parallel processing based on software-defined radio. EURASIP J. Wireless Commun. Netw 2013, 2013(1):228. doi:10.1186/1687-1499-2013-228. Accessed 26 May 2014 10.1186/1687-1499-2013-228
19. Shepard D: A two-dimensional interpolation function for irregularly-spaced data. In Proceedings of the 1968 23rd ACM National Conference. Las Vegas, NV, USA; 1968:517-524. doi:10.1145/800186.810616
20. Plets D, Joseph W, Vanhecke K, Tanghe E, Martens L: Coverage prediction and optimization algorithms for indoor environments. EURASIP J. Wireless Commun. Netw 2012, 2012(1):1-23. doi:10.1186/1687-1499-2012-123 10.1186/1687-1499-2012-1
21. Information Sciences Institute, University of Southern California: RFC 791: Internet Protocol. University of Southern California, Los Angeles; 1981.
22. ITU-T Study Group 12: ITU-T Recommendation G.107. International Telephone Connection and Circuits. General Definitions, The E model, a Computational Model for Use in Transmission Planning. 2014. International Telecommunication Union. http://handle.itu.int/11.1002/1000/12120
23. Iperf: The TCP/UDP bandwidth measurement tool. . Accessed 19 July 2013 http://iperf.fr/
24. IEEE: IEEE Std 802.11-2012: IEEE Standard for Information Technology-Telecommunications and Information Exchange Between Systems Local and Metropolitan Area Networks?Specific Requirements part 11: Wireless LAN Medium Access Control (MAC) and Physical Layer (PHY) Specifications (Revision of IEEE Std 802.11-2007). 2012, 1-2793.
25. Heusse M, Rousseau F, Berger-Sabbatel G, Duda A: Performance anomaly of 802.11b. In Twenty-Second Annual Joint Conference of the IEEE Computer and Communications. San Francisco, CA, USA; 2003:836-8432. doi:10.1109/INFCOM.2003.1208921

26. Alcock S, Lorier P, Nelson R: Libtrace: a packet capture and analysis library. ACM SIGCOMM Computer Communication Review 2012, 42-48. doi:10.1145/2185376.2185382
27. Tcpdump/Libpcap: tcpdump: a powerful command-line packet analyzer.. Accessed 19 July 2013 http://www.tcpdump.org/
28. Gamal AE, Mammen J, Prabhakar B, Shah D: Throughput-delay trade-off in wireless networks. INFOCOM 2004, 1: 475. doi:10.1109/INFCOM.2004.1354518

INDEX